Trends in Ornamental Plant Production

Trends in Ornamental Plant Production

Editor

Piotr Salachna

MDPI • Basel • Beijing • Wuhan • Barcelona • Belgrade • Manchester • Tokyo • Cluj • Tianjin

Editor
Piotr Salachna
West Pomeranian University
of Technology in Szczecin
Poland

Editorial Office
MDPI
St. Alban-Anlage 66
4052 Basel, Switzerland

This is a reprint of articles from the Special Issue published online in the open access journal *Horticulturae* (ISSN 2311-7524) (available at: https://www.mdpi.com/journal/horticulturae/special_issues/Ornamental_Plant).

For citation purposes, cite each article independently as indicated on the article page online and as indicated below:

LastName, A.A.; LastName, B.B.; LastName, C.C. Article Title. *Journal Name* **Year**, *Volume Number*, Page Range.

ISBN 978-3-0365-4685-8 (Hbk)
ISBN 978-3-0365-4686-5 (PDF)

Cover image courtesy of Piotr Salachna

Contents

About the Editor

Piotr Salachna

Professor Piotr Salachna is a lecturer of horticulture and floristry at the West Pomeranian University of Technology in Szczecin, Poland. He was born in Debica and earned his MS (2001) and PhD (2006) from the University of Agriculture in Szczecin. His main areas of expertise include the use of various natural biostimulants, biomaterials and nanoparticles in ornamental and medicinal plant production.

horticulturae

MDPI

Editorial

Trends in Ornamental Plant Production

Piotr Salachna

Department of Horticulture, West Pomeranian University of Technology, 3 Papieża Pawła VI Str., 71-459 Szczecin, Poland; piotr.salachna@zut.edu.pl

Growing ornamental plants is a dynamically developing and profitable sector of plant production. In 2019, the value of the flower market on the largest global flower exchange, Royal FloraHolland, reached 4.8 billion euros. In 2021, despite the problems related to SARS-CoV-2 and the global pandemic, the value of the annual flower trade increased to 5.6 billion euros [1]. The power of the market and the floriculture sector lies in the variety of the assortment offered for sale. This is why it is so important to constantly introduce new species and cultivars, especially those with the most environmentally friendly production process [2,3]. Other factors important for the constant expansion of the floriculture industry include implementing new strategies for plant reproduction, regulating their growth and development, adapting production technologies to fit the idea of sustainable development, and optimizing supply chain management [4,5]. All these aspects are discussed in seven papers published in this Special Issue on the 'Trends in Ornamental Plant Production'.

The modern production of ornamental plants requires solutions that combine improved production efficiency with a more rational and environmentally friendly use of resources. The principle of sustainable development is a perfect answer to these challenges, as it allows for more effective use of the means of production and better protection of the environment in which a production facility operates. One of the elements of sustainable development in floriculture is biological progress achieved by implementing the species with low thermal requirements and relatively good resistance to diseases and pests. This topic is discussed in depth in an interesting review article [6] that characterizes specialty cut flowers (SCF) and their increasingly important role in the global and local floricultural market. The SCF group is not homogeneous, and it includes annual species, biennials, perennials, bulbs, and woody plants. The main source of their genotypes is the endemic flora of South Africa, Australia, and America. In comparison with traditional cut flowers (TCF), such as roses, gerberas, or anthuriums, the production of SCF flowers is considerably less energy-consuming and safer for people and the environment. This aforementioned paper presents a SWOT analysis that comprehensively assesses the external and internal factors determining the development potential of SCF and TCF flower production. It also discusses the issues related to the harvest, storage, and extension of the vase life of little-known cut flowers.

The environmental management technique called 'life cycle assessment' (LCA) is a tool defined in ISO standards and recommended in many EU documents. It enables a comprehensive assessment of environmental hazards. LCA identifies and prioritizes individual risks and is therefore helpful in the search for technological solutions aimed at maintaining optimal environmental quality. The authors in [7] describe the use of LCA in assessing the environmental impact of *Cyclamen persicum* and *Pelargonium* ×*hortorum* production in 20 horticultural farms from the floriculture district of Treviso, Veneto region. LCA analysis showed that the production of *P.* ×*hortorum* more strongly affected the environment than that of *C. persicum*, mainly due to fossil fuel consumption for heating greenhouses. However, the production of C. persicum is more variable and affects the environment in a more diverse way, especially in the field of eutrophication, acidification, and human toxicity potential. The authors point out that growing *C. persicum* in accordance

Citation: Salachna, P. Trends in Ornamental Plant Production. *Horticulturae* **2022**, *8*, 413. https://doi.org/10.3390/horticulturae8050413

Received: 1 May 2022
Accepted: 2 May 2022
Published: 6 May 2022

with the principles of integrated pest management and using compost to significantly limit the consumption of mineral fertilizers brings about measurable benefits for the environment and human health.

Plant growth regulators (PGRs) act at very low concentrations to stimulate, inhibit or otherwise modify plant growth. Although there are many research studies on PGRs, specific responses of individual plant species and cultivars make PGR use a still up-to-date and interesting topic. In [8], an international team presents the effects of abscisic acid, N-acetyl thiazolidine, gibberellic acid, salicylic acid, indole-3-butyric acid, and oxalic acid on the flowering and antioxidant potential of *Tagetes erecta*, a popular ornamental, edible, and medicinal plant. The authors demonstrated that PGRs effectively improved the flowering and antioxidant potential of *T. erecta* inflorescences depending on the genotype and type and concentration of PGR used. In *T. erecta* cv. 'Narangi', foliar treatment with different concentrations of oxalic acid considerably enhanced inflorescence biomass, the total content of polyphenols and flavonoids, as well as antioxidant capacity. As for *T. erecta* cv. 'Basanthi', the antiradical activity of the extracts was the most strongly influenced by spraying the plants with indole-3-butyric acid at 100 mg/l.

PGRs are commonly used for root induction and development in cuttings propagated in ornamental shrub nurseries. Auxins are particularly capable of stimulating simultaneous and steady root formation. This is especially important in the intensive reproduction of species with a poor ability to form adventitious roots on cuttings. Researchers in [9] present the impact of 1-naphthylacetic acid (NAA) on rooting effectiveness in *Syringa vulgaris* cv. 'Mme Lemoine', *S. vulgaris* cv. 'President Grevy', *Ilex aquifolium*, *Cotinus coggygria*, *Cotinus coggygria* cv. 'Kanari', and *C. coggygria* cv. 'Royal Purple'. All these shrub taxa, except for *Cotinus coggygria* cv. 'Royal Purple', positively responded to NAA application. A particularly beneficial effect of NAA on the rooting percentage of the cuttings, root volume, number of roots, and root length and diameter of the cuttings was observed in *Ilex aquifolium*.

PGRs are being increasingly replaced with biostimulants to improve plant growth and quality. A valuable source of biostimulants is natural polysaccharides and their derivatives. They are safe for the environment and therefore provide a perfect alternative to synthetic PGRs. The experiments reported in [10] investigated two types of biostimulant complexes, composed of depolymerized chitosan and carrageenan and depolymerized chitosan and xanthan, and assessed their stimulating effects on the growth and quality of *Eucomis autumnalis*. *E. autumnalis* is an endemic species grown as an ornamental and medicinal plant. The biostimulants were applied using a patented method of bulb coating prior to their planting. Both biostimulant complexes effectively improved growth and flowering, increased bulb yield, shortened the period of plant production, and enhanced the content of macroelements and total sugars in *E. autumnalis*. The coating of ornamental plant bulbs in biostimulants is an environmentally friendly biostimulation method with a promising future in sustainable cultivation systems.

The production of potted plants is developing very dynamically, and the practical aspects of their cultivation are always of great importance for producers. In [11], a group of researchers presents their findings related to the effect of temperature on the growth of hydroponically cultivated *Streptocarpus formosus*. This is still a little-known but very attractive plant, native to South Africa, and recommended for cultivation in pots, flower beds, and as a cut flower. A lower root-zone temperature (18 °C) increased the leaf number, leaf and root length, and fresh weight, while a higher root-zone temperature limited vegetative growth of *S. formosus*. Increasing the root-zone temperature during the plant dormancy did not stimulate flowering.

To streamline the supply chain in the floriculture industry, researchers from Ecuador and Spain developed the SCOR (Supply Chain Operations Reference) model and a multicriteria decision-making method [12] based on questionnaires filled by companies representing this sector. The model can be used to assess the performance of individual companies as well as the performance of the entire floriculture sector. The authors concluded that Ecuadorian floriculture farms need to improve their planning, procurement, and manufacturing.

In summary, the production of ornamental plants, just as in other horticulture sectors, is subject to constant changes. The long-term development of this industry, faced with the current energy crisis, post-pandemic challenges, and threats to global geopolitical stability, is highly uncertain. Therefore, to continue its constant development, it is necessary to adapt cultivation methods to actual conditions and take into account the energy transformation and biological, technical, and organizational advances. It is also increasingly important that all flower companies systematically implement sustainable development strategies, which requires a favorable political and social atmosphere.

Acknowledgments: I am deeply grateful to the authors of the papers for sharing their research results in this Special Issue, to the reviewers for their unbiased and insightful reviews, and to the entire team of the *Horticulturae* editors I had the pleasure to work with.

Conflicts of Interest: The author declares no conflict of interest.

References

1. Royal FloraHolland Annual Report 2021. Available online: www.floraholland.com (accessed on 23 April 2022).
2. Darras, A. Implementation of Sustainable Practices to Ornamental Plant Cultivation Worldwide: A Critical Review. *Agronomy* **2020**, *10*, 1570. [CrossRef]
3. Wani, M.A.; Nazki, I.T.; Din, A.; Iqbal, S.; Wani, S.A.; Khan, F.U. Floriculture sustainability initiative: The dawn of new era. In *Sustainable Agriculture Reviews*; Springer: Cham, Switzerland, 2018; Volume 27, pp. 91–127.
4. Gabellini, S.; Scaramuzzi, S. Evolving Consumption Trends, Marketing Strategies, and Governance Settings in Ornamental Horticulture: A Grey Literature Review. *Horticulturae* **2022**, *8*, 234. [CrossRef]
5. Krigas, N.; Tsoktouridis, G.; Anestis, I.; Khabbach, A.; Libiad, M.; Megdiche-Ksouri, W.; Ghrabi-Gammar, Z.; Lamchouri, F.; Tsiripidis, I.; Tsiafouli, M.A.; et al. Exploring the Potential of Neglected Local Endemic Plants of Three Mediterranean Regions in the Ornamental Sector: Value Chain Feasibility and Readiness Timescale for Their Sustainable Exploitation. *Sustainability* **2021**, *13*, 2539. [CrossRef]
6. Darras, A. Overview of the Dynamic Role of Specialty Cut Flowers in the International Cut Flower Market. *Horticulturae* **2021**, *7*, 51. [CrossRef]
7. Bonaguro, J.E.; Coletto, L.; Sambo, P.; Nicoletto, C.; Zanin, G. Environmental Analysis of Sustainable Production Practices Applied to Cyclamen and Zonal Geranium. *Horticulturae* **2021**, *7*, 8. [CrossRef]
8. Sadique, S.; Ali, M.M.; Usman, M.; Hasan, M.U.; Yousef, A.F.; Adnan, M.; Gull, S.; Nicola, S. Effect of Foliar Supplied PGRs on Flower Growth and Antioxidant Activity of African Marigold (*Tagetes erecta* L.). *Horticulturae* **2021**, *7*, 378. [CrossRef]
9. Kentelky, E.; Jucan, D.; Cantor, M.; Szekely-Varga, Z. Efficacy of Different Concentrations of NAA on Selected Ornamental Woody Shrubs Cuttings. *Horticulturae* **2021**, *7*, 464. [CrossRef]
10. Salachna, P.; Pietrak, A. Evaluation of Carrageenan, Xanthan Gum and Depolymerized Chitosan Based Coatings for Pineapple Lily Plant Production. *Horticulturae* **2021**, *7*, 19. [CrossRef]
11. Viljoen, C.C.; Jimoh, M.O.; Laubscher, C.P. Studies of Vegetative Growth, Inflorescence Development and Eco-Dormancy Formation of Abscission Layers in *Streptocarpus formosus* (Gesneriaceae). *Horticulturae* **2021**, *7*, 120. [CrossRef]
12. Rodríguez Mañay, L.O.; Guaita-Pradas, I.; Marques-Perez, I. Measuring the Supply Chain Performance of the Floricultural Sector Using the SCOR Model and a Multicriteria Decision-Making Method. *Horticulturae* **2022**, *8*, 168. [CrossRef]

horticulturae

MDPI

Article

Measuring the Supply Chain Performance of the Floricultural Sector Using the SCOR Model and a Multicriteria Decision-Making Method

Luís Oswaldo Rodríguez Mañay [1], Inmaculada Guaita-Pradas [2] and Inmaculada Marques-Perez [2,*]

[1] Facultad de Ciencias Administrativas, Universidad Central del Ecuador, Quito 170129, Ecuador; lorodriguez@uce.edu.ec

[2] Faculty of Business Administration and Management, Economics and Social Sciences Department, Universitat Politècnica de València, 46022 Valencia, Spain; iguaita@esp.upv.es

* Correspondence: imarques@esp.upv.es

Abstract: This study aims to highlight the usefulness of studying the performance of supply chains (SC) at the sectoral level in greater detail through the combination of a disaggregated supply chain operations reference (SCOR) model, with a multicriteria decision-making approach, specifically using an AHP, to adjust the analysis to the particularities of the sector under study by stakeholders' judgements. The methodology was applied to the Ecuadorian flower industry, and the data for the analysis was from a survey of a group of companies that represent this sector. In addition, a focus group of SC experts weighted the model constructs as part of the analytic hierarchy process (AHP), and then the performance level for each construct was determined. According to the results methodologies, this model allows the classification of companies by their performance, as well as the performance of the aggregate sector. The processes that Ecuadorian flower companies need to improve on are planning, procurement, and manufacturing. The study's main contribution is developing a general framework for measuring the overall performance of SCs and how the results are obtained. This tool could help managers, consultants, industries, and governments to assess the performance of SCs, as well as improving SC management in order to increase the sector's competitiveness in the international market.

Keywords: supply chain performance; floricultural sector; SCOR; AHP

Citation: Rodríguez Mañay, L.O.; Guaita-Pradas, I.; Marques-Perez, I. Measuring the Supply Chain Performance of the Floricultural Sector Using the SCOR Model and a Multicriteria Decision-Making Method. *Horticulturae* **2022**, *8*, 168. https://doi.org/10.3390/horticulturae8020168

Academic Editor: Piotr Salachna

Received: 14 December 2021
Accepted: 9 February 2022
Published: 16 February 2022

1. Introduction

Supply chains (SC), which are understood as a system of people, organizations, activities, resources, and data that are involved in the flow of products or services from the supplier to the customer [1] have developed continuously over the past forty years [2], especially during the months following the outbreak of the pandemic [3]. SCs evolve for two reasons: (i) to improve their performance and the system's functioning, as well as the elements that make it up, and (ii) to ensure consumer satisfaction [4]. Recently, the COVID-19 pandemic has exposed the vulnerability of supply chain risk management [5]. The concept of *supply chain management* (SCM) was first introduced in the 1980s to express the need to integrate all the processes of a supply chain, from the end-user to the original suppliers [6,7]. Since then, plenty of research has been undertaken both to study supply chain management in various fields of activity (industry, transportation, distribution, and agriculture, among others), as well as to measure and determine the ability of different SC processes to achieve their set goals, and also, to identify processes which could be improved in order to make SCs more effective and efficient. Among the most recent works, the following are worth highlighting: supply chain risk management (SCRM) [8], environmental supply chain management (GSCM) [2], and supply chain performance management (SCMP) [9,10], as well as those works exploring the use of technologies, such as artificial intelligence (AI) [11,12], the Internet of Things (IoT) [12], 3D printing [12], big data [12], and blockchain [12,13].

It should also be stressed that, nowadays, supply chain management (SCM) needs to adapt to more dynamic environments characterized by competition, rapidly evolving technologies, and higher consumer expectations for responsiveness [14]. All of these circumstances put pressure on supply chains to be more integrative and collaborative [4]. SC integration enables SC systems to shorten their response time because it allows the frequent and rapid changes in markets and demand to be managed [2]. Silvestro and Lustrato [15] emphasized the importance of integrating physical supply chain activities for several reasons: (1) it provides quick responses to fast-moving markets under conditions of demand uncertainty [16]; (2) it enables a closer collaboration between buyers and sellers along the supply chain, resulting in significant reductions in delivery times and costs [17,18]; (3) an integrated supply chain works better than each process on its own [19,20]; and (4) it maximizes information visibility through the use of the Internet and the involvement of all the parties in the supply chain [21,22]. Given that information transparency along the supply chain has become a priority for buyers and suppliers, highly complex supply chain networks tend to improve their performance when integrated [13,23].

There are several models and techniques for assessing SC performance, of which the following stand out: (a) the supply chain operation reference (SCOR) model, which is a model that describes, communicates, assesses, and identifies opportunities for improving workflow efficiency [4]; (b) the Global Supply Chain Forum (GSCF) model, which provides a systematic overview of the balance, alignment, and management of SC technological capabilities to achieve successful management [4]; (c) the Triple-E model, developed by Simao et al. (2021) [10], which focuses on three performance dimensions: efficiency, efficacy, and environmental impact [10]; and (d) the BSC model, developed by Kaplan and Norton, which allows managers to obtain an overall view of a supply chain's performance [24,25].

Developed and endorsed by the Supply Chain Council (SCC, focuses primarily on defining the core processes that make up a supply chain system) as an industry-standard diagnostic tool, the SCOR model emerged in 1996 and, since then, it has evolved from its initial design to its current 12th version (The SCC with American Production and Inventory Control Society (APICS) produced the latest SCOR version, 12.0, in 2017) [10]. It is a powerful tool for structuring, assessing, and comparing supply chain practices and performance [26,27]. Furthermore, it is known to be an integrated approach based on the idea that the SC is an interconnected structure that combines SC processes, performance metrics, best practices, and technology into a single framework for the effective communication and the continuous improvement of the SC [5]. Moreover, it has been increasingly used by practitioners and academics involved in value chain management [28] and, in general, it is a global benchmark that enables the comparisons of SCs [29].

In recent years, several studies of supply chain management have combined the SCOR model with multi-criteria techniques to improve the analysis of SCs. Table 1 provides a list of these combinations, along with the works.

Table 1. Combinations of the SCOR model with multi-criteria techniques for studying supply chain management.

Authors	Methods Applied	Aim
Nisa Afifa & Santoso, 2018 [30]	SCOR–FUZZY–ANP	Supply chain risk management
Effendi et al., 2019 [31]	SCOR–DEMATEL	Assess the performance of green supply chain management
Büyüksaatçi Kiriş et al., 2019 [32]	SCOR–FUZZY DEMATEL	Evaluate suppliers' performance

Table 1. *Cont.*

Authors	Methods Applied	Aim
Wang et al., 2018 [33]	SCOR–AHP–TOPSIS	Select suppliers
Lima-Junior & Carpinetti, 2016 [34]	SCOR–FUZZY TOPSIS	Assess suppliers
Lhassan et al., 2018 [35]	SCOR–BPMN	Map supply chain processes
Teixeira & Borsato, 2019 [36]	SCOR–BPMN	Dynamic formation of supplier networks to optimize SCs
Liu et al., 2018 [37]	BSC–SCOR	Green supply chain management
Wang, Yang, et al., 2019 [38]	SCOR–FANP–TOPSIS	Select suppliers
Wang, Van Thanh, et al., 2019 [39]	SCOR–FANP–VIKOR	Select suppliers
Wang, Tsai, et al., 2020 [40]	SCOR–AHP–DEA	Select suppliers

Source: Authors' review.

In this regard, various studies have combined the analytic hierarchy process (AHP) [41] with the SCOR model in supply chain analyses [42]. The most relevant works are listed in Table 2.

Table 2. Research using both the SCOR model and AHP approaches to improve supply chain performance.

Authors	Techniques	Aim
Kocaoğlu et al., 2013 [42]	SCOR–AHP–TOPSIS	Decision-making process in a manufacturing company for the construction industry
Wang, Hoang Viet, et al., 2020 [43]	SCOR–ANP–FAHP–PROMETHEE II	Select suppliers in textile industry
Bukhori et al., 2015 [44]	SCOR–AHP–Cause Effect Diagram	Identify performance issues in poultry supply chain by a poultry company
Palma-Mendoza & Neailey, 2015 [45]	SCOR–AHP–BPR	Redesign business processes in an Airline MRO supply chain
Sellitto et al., 2015 [46]	SCOR–AHP	Measure SC performance in the Brazilian footwear industry
Sutoni et al., 2021 [47]	SCOR–AHP	P.T. performance X for the production, warehouse, and shipping of goods in a company
Nguyen et al., 2021 [48]	SCOR–AHP	Measure performance of the Vietnamese coffee supply chain
Defrizal et al., 2020 [49]	SCOR–AHP	Analyze how rice supply and rice supply chain systems work
Novar et al., 2018 [50]	SCOR–AHP	Monitor the metrics of a supply chain measurement system

Source: Authors' review.

The SCOR model is based on a hierarchical structure with four different levels. Level 1 presents the different types of processes and identifies the scope and content of the supply chain. Level 2 presents the process categories that include the operations (sub-processes), while Level 3 corresponds to the process elements that form the individual process configurations (tasks that are grouped by activities in each sub-process) [48]. The first point to consider, when analyzing the SCOR model processes, is to check which ones need to be analyzed, as well as the level of disaggregation, i.e., whether they are primary processes, sub-processes, specific activities, or tasks. In addition, it is necessary to establish a measuring system with which the values that reflect the level of performance of these processes can be calculated [51].

In general, it can be observed that previous SC assessments using the SCOR model and the associated performance metrics predominantly analyzed supply chains' main processes, but very few of them considered a division into sub-processes and activities, and almost none on them considered a disaggregation into tasks [52,53]. However, an analysis of the individual processes, sub-process, activities, and tasks could help to better identify where the problems originate in each process; in other words, it would enable us to identify which process, or sub-processes, activities, or tasks are more critical, why they are critical, what the causes are, and how they can be corrected.

This approach has been applied to the Ecuadorian flower industry.

In distributing and selling perishable products, such as flowers, supply chain management is a crucial and decisive element in improving their efficiency, productivity, and the overall distribution and sale processes. Ecuador is the third-largest producer of cut flowers in the world, where flower companies are a significant source of income and employment for this country [1]. The Ecuadorian floriculture industry is characterized by short product life cycles, a wide product variety, volatile and changing demand, and long and inflexible delivery processes [2,3]. Since 2021, due to the COVID-19 pandemic, it has also been beset by international trade and transport problems [4], which have affected the production and marketing of thousands of products traded around the world. With regard to the Ecuadorian flower sector, in particular, the greatest impact of the COVID-19 crisis has been due to a rise in the price of inputs and fertilizers [5], as well as the lack of air freight companies that could deliver floral products on time, with the required quality [5,6]. These constraints and difficulties are currently exposing the supply chain (SC) management to a variety of risks and uncertainties [7,8]. Any attempt to improve the distribution channels in the floriculture sector requires a detailed analysis of its supply chain performance. The proposed performance analysis model was applied to a set of flower companies to assess how well the supply chain was performing at the individual level, and to identify the problems. The individual values were then aggregated to establish whether the supply chain was working well in sectoral terms, and similarly, where the problems lay. Currently, the Ecuadorian flower sector does not have a methodology or model to measure the performance of the supply chain. We apply this proposal to the Ecuadorian flower industry.

The content of the manuscript is structured as follows. First, the SCOR model approach, followed by the analysis of the floriculture supply chain, is explained. Then, consultations that are carried out with the sector's companies to obtain each company's performance data is described, as well as the order of processing and aggregating the survey results to work out the individual performance values. Next, using an AHP, the performance results are interpreted and discussed by analyzing the sector's performance through the individual and aggregated results. Finally, the practical and theoretical implications of the proposed methodology are discussed, as well as the most relevant issues and suggestions for future research.

With this purpose, here, we present a methodology for examining supply chains' levels of performance at the sectoral level, combining the SCOR model, that is disaggregated to Level 4, with a multi-criteria methodology (AHP) to adjust the analysis to the specificities of the sector under study, based on stakeholders' assessments. In particular, by applying the proposed methodology, we can determine which processes are the most critical, and why, as well as the causes of performance problems and how those can be corrected.

2. Materials and Methods

Figure 1 summarizes the methodology used to analyze the Ecuadorian flower sector based on the structuring of a supply chain, as defined by the SCOR model, in combination with an AHP approach.

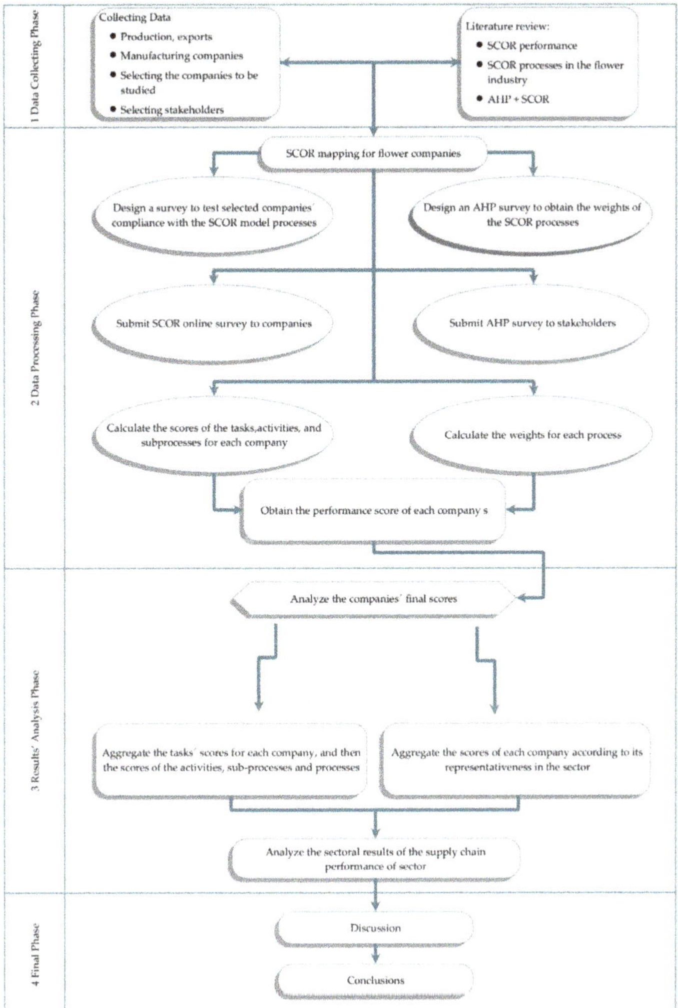

Figure 1. Methodology. Source: Authors' diagram.

As mentioned earlier, the 12th version of the SCOR model establishes a performance-analysis system with up to four levels. Thus, in addition to the first level of the supply chain, which is composed of six main processes (planning, procurement, manufacturing, distribution, return, and management), three more levels can be differentiated, namely, sub-processes, activities, and tasks, where each one might influence the main processes' performance and should, therefore, be analyzed [54].

Previous studies dealing with SC measurement, using the SCOR model, examined four, five, or six of its processes. In this study, we examined the planning, procurement, manufacturing, distribution, and return processes, which are those that are directly linked to the supply chain [55].

For example, process 1, planning, is broken down into three sub-processes [56] (see Figure 2). Each of these sub-processes is, in turn, disaggregated into different activities. For example, sub-process 1.1, supply chain planning, is decomposed into four activities, each of which is then divided into tasks (see Figure 2).

As previously pointed out, the greater the disaggregation, the better the analysis can identify the failures and where action is needed [57,58]. Our methodological proposal is to disaggregate each of the five main supply chain processes, up to level 4, which corresponds to the individual tasks.

Figure 2. *Cont.*

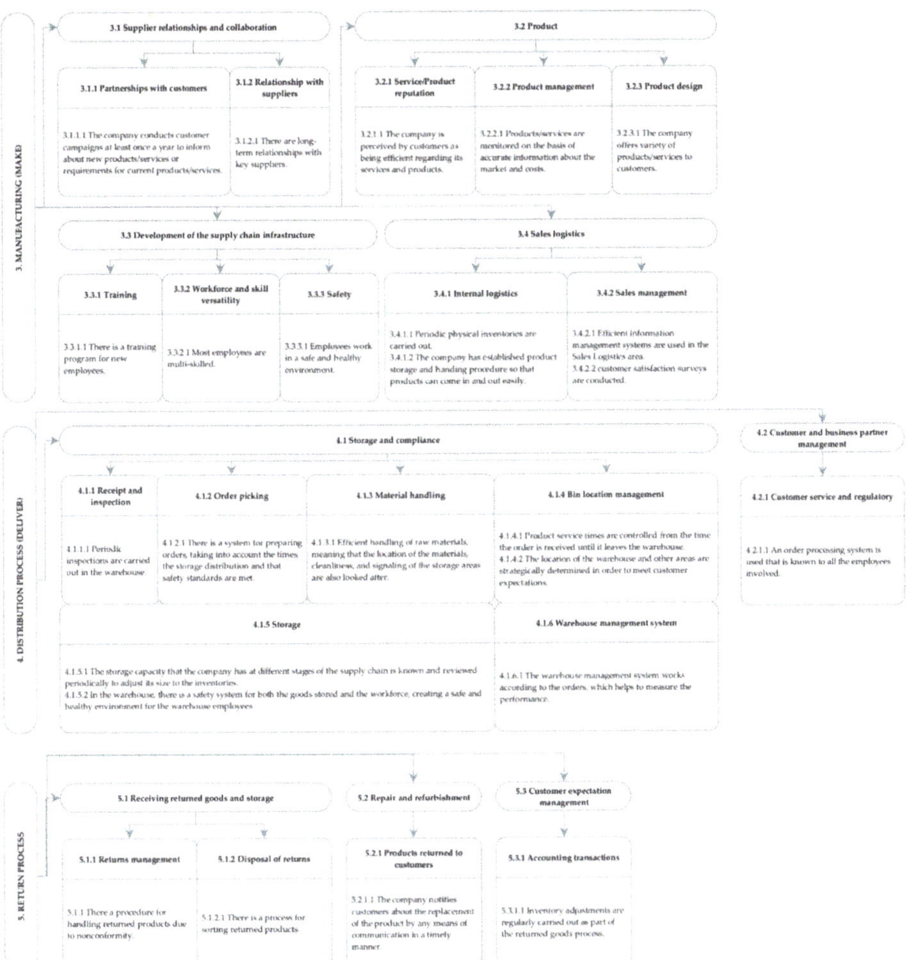

Figure 2. Disaggregation of the SCOR model's supply chain main processes into sub-processes, activities, and tasks. Source: Authors' diagram.

The proposed evaluation method is to assess all the processes and activities of the SC, in regard to their compliance with the standards. Thus, our SC performance assessment is based on checking whether the tasks, activities, sub-processes, and processes were completed or not [51]. Consequently, each company was sent a survey with a proposed breakdown of the SC and was asked to indicate whether or not it carried out the different individual tasks. A good performance involved completing each one of the defined tasks (i.e., the companies replied to dichotomous questions with yes or no answers), which meant that all activities, sub-processes, and processes were performed. After collecting the answers, we assigned a one to those tasks that were performed (for answering YES), and for those that were not carried out, a zero was assigned (for answering NO). Additionally, it is necessary to first carry out a systematic evaluation of each particular process and establish how the results are aggregated afterward to obtain a metric for measuring the SC's level of performance within the sector. To calculate the SC's overall performance, the aggregate value of the performance index must be calculated by weighting each SC process according to its relevance in the sector of activity. By using the AHP technique, it is possible to distinguish the importance of each process of the SC when aggregating the data. This distinction is made by the sector's stakeholders, based on the importance they attach to every process or activity, since the aim is to provide a metric that considers the particularities of each sector. To aggregate the single values obtained from the rating of the tasks, activities, sub-processes, and processes, Aliaga Rota et al. [51] proposed using the average of the separate scores given for each sub-process, which, in turn, are gathered from the average of the scores obtained by the activities involved in that sub-process, and so on [51,59]. The aggregation of the results is carried out, considering the importance given to each process by the stakeholders participating in the AHP [60]. As a result of this aggregation, the SC performance of each company in the sector can be analyzed.

The AHP is a technique by which experts in a given field make pairwise comparisons in order to derive priority scales. Furthermore, it provides an algorithm to solve complex decision-making problems that are broken down into a hierarchy [61,62]. This method involves two main steps [63]. First, each stakeholder completes a pairwise comparison survey, which is designed based on the hierarchy previously established, indicating which of the two elements that are compared they consider to be more important and, using Saaty's scale (Table 3), how much more important they are.

Table 3. Saaty's scale.

Intensity Scale of Importance	Definition
1	Equal importance
3	Moderate importance
5	Strong importance
7	Very strong importance
9	Extreme importance

Source: Leal [63].

The second stage of the AHP is to calculate the vector of priorities, according to the following formula:

$$p_{rj} = \frac{1}{a_{ij} * \sum_{k=1}^{n} \frac{1}{a_{ik}}} \quad (1)$$

where j is the element for which the priority is calculated, i is the base element for the comparison, a_{ij} is the value of the alternative i that is compared with the alternative j, by the criteri k, a_{ik} is the value of alternative i for the criteria k, p_{rj} is the priority of the alternative j against the considered criterion, and n is the number of criteria.

The coherence of the preferences of stakeholders was studied based on the "consistency", which should be taken into account in order to consider whether opinions are valid for determining the priorities. The consistency analysis requires calculating the "consistency index" (*CI*) of Saaty's Scale for each preferences matrix.

$$CI = \frac{\lambda_{max} - n}{n - 1} \quad (2)$$

The "consistency ratio" (*CR*) is calculated from the *CI*. The *CR* is a ratio of the *CI* and *RI*:

$$CR = \frac{CI}{RI} \quad (3)$$

where the *RI* is the average value of the *CI* of pair-wise comparisons matrices of the same order, randomly obtained. When the *CR* is less than 10% (0.1), the matrix is considered to offer acceptable consistency. Saaty's scale calculated the random indices of the RI for different matrix sizes to obtain *CR*.

There are two possibilities, when aggregating results, to analyze the company performance at the sectoral level. The first one is to aggregate the individual results obtained from individual analyses. It is then necessary to determine how the individual values will be aggregated, so that they can be interpreted in sectoral terms. The results may be aggregated by the company type, but another way is to aggregate them according to the tasks performed by all of the companies. This way, the sector performance indicator for each task is calculated; the aggregation of these indicators will result in a sector performance indicator for each activity. By aggregating these, we can then calculate the performance indicators of the sub-processes. Finally, by aggregating the latter, we can establish the level of performance of each primary process. Regardless of the method, the representativeness of each company in the sector should be considered when aggregating the individual data. This representativeness can be determined by the company's turnover. Nevertheless, in both cases, the aggregation of the processes must be carried out in consideration of the importance of each process that is given by the stakeholders, who are participating in the AHP.

Once the ratings of the five main processes have been obtained and aggregated according to the weights defined by the AHP, the overall performance score of the SC can be achieved. Table 4 contains Kusrini et al.'s [64] proposal for rating supply chains' performance, according to a scale. The scale can be used to rate each company's performance and that of the sector, as well as the disaggregated results of the tasks, activities, processes, and sub-processes.

Table 4. Performance scale.

Performance Values	Performance Indicator
<40	Poor (P)
40–50	Marginal (M)
50–70	Average (A)
70–90	Good (G)
>90	Excellent (E)

Source: Kusrini et al. [64].

3. Case Study

We tested the proposed methodology in a case study of the Ecuadorian flower industry. Ecuador is currently the third-largest exporter of cut flowers worldwide. Although Ecuador had increased its exports up until 2019, it did so at a far lower rate in both value and volume than other flower-exporting countries. The subsequent fall became more marked in 2020, due to the restrictions brought on by the pandemic [65].

Flower production has, historically, been concentrated in the provinces of Pichincha, with 62% of the production, and Cotopaxi, with 21% of production. The rest of the country's provinces, including Guayas, Imbabura, and Azuay account for the remaining 17% [66]. Furthermore, it should be noted that the industry is presently in the midst of a wave of acquisitions. In the first quarter of 2021, the largest flower company in Ecuador (Hilsea Investments, with annual sales of around USD 50,000,000) was transferred to the investment company Sunshine Bouquet, which belongs to a group of the 500 largest companies in Colombia. Additionally, a number of other small firms, namely, Alma Roses, Sisapamba, Natuflor, Romaverde, Bellarosa, Rose Connection, Qualisa, and Florasani were taken over by the investment company Elite, one of the 500 largest companies in Ecuador [67].

For the case study, we selected a representative sample of floricultural companies from the Expoflores directory, where data was accessible. Specifically, the first 96 Ecuadorian flower companies (Order established according to the income data published by the Superintendencia de Compañías del Ecuador, https://www.supercias.gob.ec/portalscvs/, accessed on 30 April 2021) were chosen. According to the value of sales, these represented approximately 70% of the more than USD 800,000,000 turnover of the sector in 2019 [68]. As seen in Table 5, the turnover in these 96 companies varies from the largest to the smallest, i.e., from USD 12,000 to USD 47,000,000. The highest concentration of companies corresponds to those with a turnover of between USD 12,000 and USD 13,500,000. This group accounts for 93% of the total turnover in the industry.

Table 5. Frequency distribution by turnover (USD).

Group	Lowest Turnover (USD)	Highest Turnover (USD)	Surveys Sent	Weight in the Whole Sector	Responses	Weight in the Sample	Participation Rate
1	12,000	6,742,000	66	69%	19	66%	29%
2	6,742,000	13,472,000	23	24%	9	31%	39%
3	13,472,000	20,202,000	4	4%		0%	0%
4	20,202,000	26,932,000	2	2%	1	3%	50%
5	26,932,000	33,662,000	0	0%		0%	
6	33,662,000	40,392,000	0	0%		0%	
7	40,392,000	47,122,000	1	1%		0%	0%
			96	100%	29	100%	100%

Source: Authors' calculations.

Of the 96 companies to which we sent the survey, 29 answered. Table 5 shows the results of the frequency distribution analysis of the companies that answered the questionnaire. This analysis was performed to verify how representative they are. The frequencies were calculated according to the firms' turnovers. Most of the companies in the sample that answered were from the groups with the largest number of flower companies. Table 5 shows that the weight of the companies in the sample is similar to the weight of all companies in the Ecuadorian floriculture industrial sector in each turnover group.

We used a digital questionnaire (https://docs.google.com/forms/d/1GZDfiJLW5D7IdsgrpjbXI696UlHAmH5tEOGQmmk-RKc/edit, accessed on 3 June 2020) to collect the preferences. Although various alternatives were available, we chose Google forms for this study. The form was sent to the companies' representatives by email, along with a letter explaining the study's purpose: to analyze the supply chain of Ecuador's flower industry, identify the key problems, and improve certain aspects.

The questionnaire was divided into four sections. The first section described the objective of the study and the survey and asked for the company's details. It also provided information on the Ecuadorian flower industry and the SC processes, as defined by the SCOR model. The following sections contained the questions about the supply chain processes. These were broken down to task levels. Respondents had to indicate which sub-processes, activities, and tasks they performed for each process.

Twenty-nine companies answered the survey, which accounted for approximately 20% of the total turnover of the selected sample, i.e., USD 180,000,000. Falcon Farms is the second-largest flower company, in terms of turnover, within this group of companies. The tasks were graded according to their fulfillment: a positive answer scored a one, and a negative answer scored a zero. Next, the average of the scores obtained for each sub-process was calculated, and then the average of the processes' scores was calculated for each of the 29 companies that answered the survey [51,69].

The importance of the SCOR model processes was determined by a group of stakeholders in the Ecuadorian floriculture sector by means of an AHP model. For this purpose, an online survey was undertaken. It was assumed that all members of the group had the same level of importance in the decision-making processes [70]. The stakeholders were: representatives of floriculture companies (6), supply chain teachers (2), experts in floriculture issues (1), and experts in quality control (1). The AHP methodology was applied to calculate the weights of the Level 1 metrics and the attributes of the SCOR model. A questionnaire was carried out that was divided into four sections. The first section described the study's objective and that of the questionnaire and requested information on the company or institution's identity. Furthermore, it included information on the Ecuadorian flower sector and descriptions of the performance attributes of the Ecuadorian supply chain, as well as the AHP hierarchy, with the objective of redesigning its elements, metrics, and processes, and an explanation of Saaty's scale for making the comparisons. The second section listed the questions related to the pairwise comparisons of the supply chain processes' attributes, in order to determine their importance (10 questions). The third section presented questions regarding the importance of the metrics for each attribute (7 questions). Finally, in the fourth section, the questions about the relevance of the performance metrics to the supply chain processes were included (10 questions).

After collecting the preferences of the stakeholders by the processes considered, we aggregated the preferences of individuals and obtained the preferences matrix from stakeholders. This was used to calculate stakeholders' priorities. Table 6 shows the weights given by the stakeholders.

Table 6. Weighted results by process.

No.	Process	Weight
1	Planning	0.4051
2	Procurement	0.1986
3	Manufacturing	0.1735
4	Distribution	0.1381
5	Return	0.0847

Source: Authors' calculations.

We have a consistency ratio of $CR = 0.0209 \leq 0.10$, so the data comparing the main criteria pairs is appropriate and does not need to be re-evaluated.

Once the weights of the processes were calculated, the scores for each company were computed according to the results of the survey.

4. Results

We determined the supply chain performance level for each of the 29 companies that answered the questionnaire using the survey data. Then, the individual results were aggregated to determine the level of performance of the supply chain at the sector level. According to the analysis and the classification proposed by Kusrini [64], the 29 firms showed a good overall performance (see Tables 4 and 7). This rating was obtained because the score achieved by each of the five processes of the SCs of the companies studied was rated as "good" (G). However, it should be noted that, as the scores obtained for the processes were less than one, all procedures need to be reconfigured and improved.

Table 7. Calculation of the sector-level performance metrics.

Process	AHP Weight	Average	Performance	Performance Metric
Planning	0.4051	0.86	G	0.35
Procurement	0.1986	0.88	G	0.17
Manufacturing	0.1735	0.79	G	0.14
Distribution	0.1381	0.88	G	0.12
Return	0.0847	0.80	G	0.07
	1.00			0.85

Source: Authors' calculations.

The turnover of the 29 companies showed a relatively low correlation (0.08) with the SC performance index, which means that the supply chain performance does not explain the sales behavior.

When considering each of the SCOR processes at the sector level, it should be highlighted that the processes with the highest GAPs (GAP: gap or difference between the intended result and the actual result obtained by the research), weighted according to their weight, were planning (0.06) and manufacturing (0.04) (see Table 8).

Table 8. Supply chain performance GAPs at the sector level by process.

Process	AHP Weight	Performance Metric	GAP
Planning	0.4051	0.35	0.06
Procurement	0.1986	0.17	0.02
Manufacturing	0.1735	0.14	0.04
Distribution	0.1381	0.12	0.02
Return	0.0847	0.07	0.02
	1.00	0.85	0.15

Source: Authors' calculations.

To improve our analysis results and to better identify where the most critical points of the SC are, we also examined the sub-processes.

Regarding the analysis of the sub-processes, of the 16 sub-processes examined (shown in Figure 2), four were rated as "excellent" (E), eleven as "good" (G) and one as "average" (A) (see Table 9). Hence, the floriculture sector should pay attention to the sub-processes with "good" and "average" ratings.

Table 9. Supply chain performance GAPs at the sector level by sub-process.

Code	Sub-Processes	AHP Weight	Performance Metric	Performance	GAP
1	**Planning process (PLAN)**	41%			
1.1	Supply chain planning		80%	G	20%
1.2	Linearity of the supply chain (alignment of supply and demand)		85%	G	15%
1.3	Inventory management		92%	E	8%
2	**Procurement process (SOURCE)**	20%			
2.1	Strategic sourcing		92%	E	8%
2.2	Supplier management		89%	G	11%
2.3	Buying products and services		82%	G	18%
2.4	Management of inbound logistics		89%	G	11%
3	**Manufacturing process (MAKE)**	17%			
3.1	Supplier relationships and collaboration		72%	G	28%
3.2	Product		92%	G	8%
3.3	Development of the supply chain infrastructure		74%	G	26%
3.4	Sales logistics		77%	G	23%
4	**Distribution process (DELIVER)**	14%			
4.1	Storage and compliance		90%	E	10%
4.2	Customer and business partner management		86%	G	14%
5	**Return process**	8%			
5.1	Receiving returned goods and storage		78%	G	22%
5.2	Repair and refurbishment		93%	E	7%
5.3	Customer expectation management		69%	A	31%

Source: Authors' calculations.

By evaluating the different activities of each sub-process, we assigned each activity the corresponding Kusrini rating. As a result, several activities with "good," "average," and "marginal" ratings need to be improved. The most critical activities, which received the lowest ratings, are:

(1) Activities with a "marginal" rating:
    - One-to-one (task) training, i.e., there is a training program for new employees (45%).

(2) Activities with an "average" rating:
    - Methods for estimating needs related to the task, i.e., statistical techniques are used to estimate the needs and validate the data sources employed to make these estimates (59%);
    - The authorization of casual purchases related to each task, i.e., casual purchases that do not exceed a certain amount, as defined by the company, are authorized (66%);
    - Feedback from customers concerning each task, i.e., the company undertakes customer satisfaction surveys at least once a year (52%);
    - Workforce and skill versatility, i.e., workers regularly switch jobs since they know how to do them (66%);
    - Sales management related to each task, i.e., the company undertakes customer satisfaction surveys (62%);
    - Returned goods management, i.e., there is a system for classifying returned goods (69%);
    - Accounting transactions, i.e., inventory adjustments are regularly carried out as part of the returned goods process (69%).

By examining the results at the company level, we can determine which companies in the sector are having the greatest problems and, therefore, need to optimize their processes. It also enables us to see which processes in each company are performing poorly. The analysis at the company level can be done individually, or by groups of companies. Table 10 shows that no company received a "poor" or "marginal" rating; four companies were rated with "average" performances, thirteen companies gave a "good" performance, and twelve gave an "excellent" performance.

Table 10. Summary of the SC performance metrics for the 29 companies that answered the survey.

Performance Values	Performance Indicator	No. of Companies	Performance Metric
<40	Poor	0	
40–50	Marginal	0	
50–70	Average	4	0.60
70–90	Good	13	0.84
>90	Excellent	12	0.94
	Total	29	0.85

Source: Authors' calculations.

Together, the four flower companies with an "average" performance rating (see Table 10) achieved a performance level of 60%. Their turnover ranged from USD 12,374 to USD 2,400,000 during 2012–2019. The process with the highest GAP was planning in the four companies, at 0.19; the remaining processes showed similar GAPs, close to 0.05 (see Table 11).

Table 11. Supply chain performance GAPs of groups of companies that answered the survey by index performance.

		4 Companies between 50 and 70		13 Companies between 70 and 90		12 Companies Higher Than 90	
Process	**AHP Weight**	**Performance Metric**	**GAP**	**Performance Metric**	**GAP**	**Performance Metric**	**GAP**
Planning	0.4051	0.21	0.19	0.35	0.05	0.39	0.01
Procurement	0.1986	0.14	0.06	0.18	0.02	0.18	0.01
Manufacturing	0.1735	0.12	0.06	0.13	0.04	0.15	0.03
Distribution	0.1381	0.10	0.04	0.11	0.02	0.14	0.00
Return	0.0847	0.04	0.05	0.06	0.02	0.08	0.00
	1.00	0.60	0.40	0.84	0.16	0.94	0.06

Source: Authors' calculations.

The thirteen floriculture companies that achieved a "good" performance rating have a turnover ranging from USD 118,000 to USD 26,400,000 during 2012-2019. All together, they achieved a performance of 84%. Planning and manufacturing stand out in these companies as the processes with the highest GAPs (see Table 11).

The minimum turnover of the twelve flower companies that achieved an "excellent" rating was USD 636,000, and the maximum turnover was USD 9,500,000 in the 2012–2019 period. Together, the companies achieved a performance level of 94% (see Table 10), which can be considered as "excellent."

The process that had the most GAPs was the manufacturing process, whereas the distribution and return processes did not show any GAPs (see Table 11).

Regarding the sub-processes' performance, in the group of companies (4) with average performances, two sub-processes obtained a "poor" rating, eight sub-processes obtained an "average" rating, and six sub-processes obtained a "good" rating. Therefore, no sub-processes achieved an "excellent" rating in this group.

The sub-processes carried out by the group of companies with the lowest scores (i.e., "poor") were supply chain planning (38%) and customer expectation management (25%). The following sub-processes received an "average" rating: the linearity of the supply chain (the alignment of supply and demand) (50%), inventory management (69%), strategic sourcing (65%), buying products and services (63%), the development of the supply chain infrastructure (63%), sales logistics (58%), receiving returned goods and storage (50%), and repair and refurbishment (50%). Those considered to have a "good" performance were supplier management (75%), the management of inbound logistics (70%), supplier relationships and collaboration (75%), the product (75%), storage and compliance (70%), and customer and business partner management (75%).

In the group of the companies (13) that achieved a good performance level, the analysis by sub-processes resulted in five sub-processes with an "excellent" rating, eight with a "good" one, and three with an "average" rating.

In this group, the sub-processes with an "excellent" rating were inventory management (96%), strategic sourcing (95%), supplier management (90%), the product (92%), and repair and refurbishment (100%). Those with a "good" rating were supply chain planning (81%), the linearity of the supply chain (the alignment of supply and demand) (83%), buying products and services (85%), the management of inbound logistics (88%), the development of the supply chain infrastructure (73%), sales logistics (74%), storage and compliance (89%), and customer and business partner management (77%), while those with average scores were supplier relationships and collaboration (69%), receiving returned goods and storage (65%), and customer expectation management (62%).

In relation to the group of companies rated as "excellent", the analysis of sub-processes resulted in twelve sub-processes with an "excellent" rating, and four with a "good" rating.

Regarding the third SCOR level, of the 48 activities studied, 24 achieved an "excellent" rating, 16 activities showed a good performance level, seven activities exhibited an average level, and one activity was considered to have a "marginal" performance level. Thus, the activities that the floriculture sector should pay more attention to are those with "good, average, and marginal" ratings, which accounted for 51% of the studied activities.

5. Discussion

Our proposed methodology shows that it is possible to analyze the performance of the supply chain at the sectoral level by applying the SCOR model and the AHP in a representative sample of companies in the sector. In previous research, these analyses were more limited. The majority did not disaggregate the SCOR model, and only studied the first level, regarding the processes [44,46–49]. Other studies were on unique companies and the results cannot be viewed as sectoral results [42,46,47]. There are some studies where the proposed methodology only studied a stage in the supply chain, and only one element in this stage. For example, Wang et al. [43] applied the model to a raw material supplier. Other works analyzed the sector and does not use company data. These used focus groups or stakeholders' opinions instead [46]. Sutoni et al. [47] used observations, interviews, literature reviews, and information or dates, but these were from a single company.

In general, an analysis of the individual processes, sub-process, activities, and tasks would enable us to identify which process, sub-processes, activities, or tasks, are more critical why they are more critical, what the causes are, and hence, how they can be corrected. The methodology proposed makes possible this analysis at the individual level, for each company, and at the sectoral level.

The proposed SC performance analysis method can be used with any company and with any industry, since it allows the evaluation of groups of companies that make up an industry or represent it, by aggregating the individual values. Additionally, it is a tool that determines where problems lie and their causes. It also helps to increase the competitiveness of firms and industries, and achieves long-term goals by supporting company managers, governments, policymakers, and every industry in the design of policies and measures to fix issues. Managers can use the results to benchmark their company's competitiveness and performance against other companies in their sector, or in sectors with similar characteristics. In the policy field, sector-level analyses can be used for planning purposes.

6. Conclusions

This study contributes to the current literature with a methodological proposal that uses the SCOR model, combined with an analytical hierarchy process (AHP) to measure the performance of supply chains within a given sector. We applied this methodology to individual flower companies to assess the degree of compliance of their supply chain (SC) processes and activities, with the standards set by the SCOR model regarding SC performance. In addition, we determined which tasks or activities in each company were not carried out and traced the origin of potential problems back to specific SC sub-processes, which should be checked. Moreover, the aggregation of performance data at the individual level enabled us to assess the performance at the sector level.

Here, we employed the proposed methodology to identify, calculate, and handle potential SC performance issues in the Ecuadorian floriculture industry. By conducting an in-depth study of Ecuadorian flower companies, we have been able to draw a comprehensive picture of this industry.

Based on the results for the 29 companies that answered the survey, the SC performance of the Ecuadorian flower sector is 85%. The results showed that all processes need to be improved, especially the planning and manufacturing processes. When analyzing the flower companies by groups according to their rating, the planning, procurement, and manufacturing processes with an "average" rating (50–70) showed large performance GAPs. Meanwhile, the planning and manufacturing processes of companies with a score of 70–90, which is considered "good", had the largest performance GAPs. Moreover, within the group of companies with a performance score that was higher than 90, the manufacturing process is the most critical.

Therefore, Ecuadorian flower companies should work on the first five SCOR processes, applying the standards suggested in the model. To excel, they should work on all processes, which also depend on external factors, in order to improve the flower industry's supply chain.

When conducting studies such as this one, the sample must be as representative of the industry as possible. Therefore, in general, obtaining a high response rate allows for a better analysis and results that reflect the realities of the sector. Hence, the scope of future studies about the Ecuadorian flower industry must be expanded to include a larger number of companies and a broader field of analysis, considering performance attributes such as reliability in compliance, the speed of responses, agility, costs, and the efficient management of assets and their components.

Author Contributions: Conceptualization, I.M.-P. and I.G.-P.; methodology, I.M.-P.; software, L.O.R.M.; validation, I.M.-P. and I.G.-P.; formal analysis, L.O.R.M.; investigation, L.O.R.M.; resources, L.O.R.M.; data curation, L.O.R.M.; writing—original draft preparation, L.O.R.M.; writing—review and editing, I.M.-P. and I.G.-P.; visualization, I.G.-P.; supervision, I.M.-P.; project administration, I.M.-P. All authors have read and agreed to the published version of the manuscript.

Funding: This research received no external funding.

Institutional Review Board Statement: Not applicable.

Informed Consent Statement: Not applicable.

Data Availability Statement: I Superintendencia de Compañías del Ecuador (https://www.supercias.gob.ec/portalscvs/ accessed on 30 April 2021).

Conflicts of Interest: The authors declare no conflict of interest.

References

1. Zeng, L.; Liu, S.Q.; Kozan, E.; Corry, P.; Masoud, M. A comprehensive interdisciplinary review of mine supply. *Resour. Policy* **2021**, *74*, 102274. [CrossRef]
2. Shen, W.; Hu, D.; Günay, E.E.; Kreme, G.E.O. Evolution of supply chain management: A sustainability focused review. *Int. J. Sustain. Manuf.* **2020**, *4*, 319–335. [CrossRef]
3. Moosavi, J.; Hosseini, S. Simulation-based assessment of supply chain resilience with consideration of recovery strategies in the COVID-19 pandemic context. *Comput. Ind. Eng.* **2021**, *160*, 107593. [CrossRef] [PubMed]
4. Dos Santos, T.F.; Leite, M.S.A. Performance Measurement System Based on Supply Chain Operations Reference Model: Review and Proposal. In *Contemporary Issues and Research in Operations Management*; IntechOpen: Rijeka, Croatia, 2018; pp. 29–49.
5. Phadi, N.P.; Das, S. The Rise and Fall of the SCOR Model: What after the Pandemic? In *Computational Management. Modeling and Optimization in Science and Technologies*; Patnaik, S., Tajeddini, K., Jain, V., Eds.; Springer: Cham, Switzerland, 2021; pp. 253–273.
6. Lee, S.-Y. Sustainable Supply Chain Management, Digital-Based Supply Chain Integration, and Firm Performance: A Cross-Country Empirical Comparison between South Korea and Vietnam. *Sustainability* **2021**, *13*, 7315. [CrossRef]
7. Liu, A.; Liu, H.; Gu, J. Linking business model design and operational performance: The mediating role of supply chain integration. *Ind. Mark. Manag.* **2021**, *96*, 60–70. [CrossRef]
8. Trenggonowati, D.L.; Ulfah, M.; Arina, F.; Lutfiah, C. Analysis and strategy of supply chain risk mitigation using fuzzy failure mode and effect analysis (fuzzy fmea) and fuzzy analytical hierarchy process (fuzzy ahp). *IOP Conf. Ser. Mater. Sci. Eng.* **2020**, *909*, 012085. [CrossRef]
9. Indah, P.N.; Setiawan, R.F.; Hendrarini, H.; Yektingsih, E.; Sunarsono, R.F. Agriculture supply chain performance and added value of cocoa a study in Kare Village, Indonesia. *Bulg. J. Agric. Sci.* **2021**, *27*, 487–497.
10. Simão, L.E.; Somensi, K.; Dávalos, R.V.; Rodriguez, C.M.T. Measuring supply chain performance: The triple E model. *Int. J. Product. Perform. Manag.* **2021**. [CrossRef]
11. Riahi, Y.; Saikouk, T.; Gunasekaran, A.; Badraoui, I. Artificial intelligence applications in supply chain: A descriptive bibliometric analysis and future research directions. *Expert Syst. Appl.* **2021**, *173*, 114702. [CrossRef]
12. Sundarakani, B.; Ajaykumar, A.; Gunasekaran, A. Big data driven supply chain design and applications for blockchain: An action research using case study approach. *Omega* **2021**, *102*, 102452. [CrossRef]
13. Hong, L.; Hales, D.N. Blockchain performance in supply chain management: Application in blockchain integration companies. *Ind. Manag. Data Syst.* **2021**, *121*, 1969–1996. [CrossRef]
14. MacCarthy, B.L.; Blome, C.; Olhager, J.; Srai, J.S.; Zhao, X. Supply chain evolution—Theory, concepts and science. *Int. J. Oper. Prod. Manag.* **2016**, *36*, 1696–1718. [CrossRef]

15. Silvestro, R.; Lustrato, P. Integrating financial and physical supply chains: The role of banks in enabling supply chain integration. *Int. J. Oper. Prod. Manag.* **2014**, *34*, 298–324. [CrossRef]
16. Mason-Jones, R.; Naylor, B.; Towill, D.R. Lean, agile or leagile? Matching your supply chain to the marketplace. *Int. J. Prod. Res.* **2000**, *38*, 4061–4070. [CrossRef]
17. Dan, B.; Zhang, H.; Zhang, X.; Guan, Z.; Zhang, S. Should an online manufacturer partner with a competing or noncompeting retailer for physical showrooms? *Int. Trans. Oper. Res.* **2020**, *28*, 2691–2714. [CrossRef]
18. Diehlmann, F.; Lüttenberg, M.; Verdonck, L.; Wiens, M.; Zienau, A.; Schultmann, F. Public-private collaborations in emergency logistics: A framework based on logistical and game-theoretical concepts. *Saf. Sci.* **2021**, *141*, 105301. [CrossRef]
19. Wei, S.; Yin, J.; Chen, X. Paradox of Supply Chain Integration and Firm Performance: The Moderating Roles of Distributive and Procedural Justice. *Decis. Sci.* **2020**, *52*, 78–108. [CrossRef]
20. Zhao, X.; Wang, P.; Pal, R. The effects of agro-food supply chain integration on product quality and financial performance: Evidence from Chinese agro-food processing business. *Int. J. Prod. Econ.* **2020**, *231*, 107832. [CrossRef]
21. Roy, V. Contrasting supply chain traceability and supply chain visibility: Are they interchangeable? *Int. J. Logist. Manag.* **2021**, *32*, 942–972. [CrossRef]
22. Saqib, Z.A.; Zhang, Q. Impact of sustainable practices on sustainable performance: The moderating role of supply chain visibility. *J. Manuf. Technol. Manag.* **2021**, *32*, 1421–1443. [CrossRef]
23. Khandelwal, C.; Singhal, M.; Gaurav, G.; Dangayach, G.S.; Meena, M.L. Agriculture Supply Chain Management: A Review (2010–2020). *Mater. Today Proc.* **2021**, *47*, 3144–3153. [CrossRef]
24. Balaji, M.; Dinesh, S.; Kumar, P.M.; Ram, K.H. Balanced Scorecard approach in deducing supply chain performance. *Mater. Today Proc.* **2021**, *47*, 5217–5222. [CrossRef]
25. Leksono, E.B.; Suparno, S.; Vanany, I. Integration of a Balanced Scorecard, DEMATEL, and ANP for Measuring the Performance of a Sustainable Healthcare Supply Chain. *Sustainability* **2019**, *11*, 3626. [CrossRef]
26. Mishra, R.; Singh, R.K.; Subramanian, N. Impact of disruptions in agri-food supply chain due to COVID-19 pandemic: Contextualised resilience framework to achieve operational excellence. *Int. J. Logist. Manag.* 2021, *ahead of printing*. [CrossRef]
27. Xu, Y.; Li, J.-P.; Chu, C.-C.; Dinca, G. Impact of COVID-19 on transportation and logistics: A case of China. *Econ. Res.* **2021**, 1–19. [CrossRef]
28. Es-Satty, A.; Lemghari, R.; Okar, C. Supply Chain Digitalization Overview SCOR model implication. In Proceedings of the 13th International Colloquium of Logistics and Supply Chain Management (LOGISTIQUA), Fez, Morocco, 2–4 December 2020; pp. 1–7. [CrossRef]
29. Lemghari, R.; Okar, C.; Sarsri, D. Benefits and limitations of the SCOR®model in Automotive Industries. *MATEC Web Conf.* **2018**, *200*, 19. [CrossRef]
30. Afifa, Y.N.; Santoso, I. Risk Analysis and Mitigation Using SCOR -Fuzzy ANP. *Indian J. Sci. Technol.* **2018**, *11*, 1–13. [CrossRef]
31. Effendi, U.; Dewi, C.F.; Mustaniroh, S.A. Evaluation of supply chain performance with green supply chain management approach (GSCM) using SCOR and DEMATEL method (case study of PG Krebet Baru Malang). In Proceedings of the International Conference on Green Agro-industry and Bioeconomy, East Java, Indonesia, 18–20 September 2018; IOP Conference Series: Earth and Environmental Science. IOP Publishing Ltd.: Bristol, UK, 2019; Volume 230.
32. Kiriş, S.B.; Börekçi, D.Y.; Koç, T. A Methodology Proposal for Supplier Performance Evaluation: Fuzzy DEMATEL Method with Sustainability Integrated SCOR Model. In *Advances in Intelligent Systems and Computing*; Springer: Cham, Switzerland, 2019; pp. 488–496.
33. Wang, C.-N.; Huang, Y.-F.; Cheng, I.-F.; Nguyen, V.T. A Multi-Criteria Decision-Making (MCDM) Approach Using Hybrid SCOR Metrics, AHP, and TOPSIS for Supplier Evaluation and Selection in the Gas and Oil Industry. *Processes* **2018**, *6*, 252. [CrossRef]
34. Lima-Junior, F.R.; Carpinetti, L. Combining SCOR® model and fuzzy TOPSIS for supplier evaluation and management. *Int. J. Prod. Econ.* **2016**, *174*, 128–141. [CrossRef]
35. Lhassan, E.; Ali, R.; Majda, F. Combining SCOR and BPMN to support supply chain decision-making of the pharmaceutical wholesaler-distributors. In Proceedings of the 2018 4th International Conference on Logistics Operations Management, Le Havre, France, 10–12 April 2018; pp. 1–10.
36. Teixeira, K.C.; Borsato, M. Development of a model for the dynamic formation of supplier networks. *J. Ind. Inf. Integr.* **2019**, *15*, 161–173. [CrossRef]
37. Liu, Y.; Xu, J.; Xu, M. Green Construction Supply Chain Performance Evaluation Based on BSC-SCOR. In Proceedings of the 2018 15th International Conference on Service Systems and Service Management (ICSSSM), Hangzhou, China, 21–22 July 2018; pp. 1–6.
38. Wang, C.N.; Yang, C.Y.; Cheng, H.C. Fuzzy multi criteria decision-making model for supplier evaluation and selection in a wind power plant project. *Mathematics* **2019**, *7*, 417. [CrossRef]
39. Wang, C.N.; Nguyen, V.T.; Chyou, J.T.; Lin, T.F.; Nguyen, T.N. Fuzzy Multicriteria Decision-Making Model (MCDM) for Raw Materials Supplier Selection in Plastics Industry. *Mathematics* **2019**, *7*, 981. [CrossRef]
40. Wang, C.-N.; Tsai, H.-T.; Ho, T.-P.; Nguyen, V.-T.; Huang, Y.-F. Multi-Criteria Decision Making (MCDM) Model for Supplier Evaluation and Selection for Oil Production Projects in Vietnam. *Processes* **2020**, *8*, 134. [CrossRef]
41. Saaty, T.; Vargas, L. *Models, Methods, Concepts & Applications of the Analytic Hierarchy Process*; Springer: Boston, MA, USA, 2012.
42. Kocaoğlu, B.; Gülsün, B.; Tanyaş, M. A SCOR based approach for measuring a benchmarkable supply chain performance. *J. Intell. Manuf.* **2013**, *24*, 113–132. [CrossRef]

43. Wang, C.-N.; Viet, V.T.H.; Ho, T.P.; Nguyen, V.T.; Nguyen, V.T. Multi-Criteria Decision Model for the Selection of Suppliers in the Textile Industry. *Symmetry* **2020**, *12*, 979. [CrossRef]
44. Bukhori, I.B.; Widodo, K.H.; Ismoyowati, D. Evaluation of Poultry Supply Chain Performance in XYZ Slaughtering House Yogyakarta Using SCOR and AHP Method. *Agric. Agric. Sci. Procedia* **2015**, *3*, 221–225. [CrossRef]
45. Palma-Mendoza, J.A.; Neailey, K. A business process re-design methodology to support supply chain integration: Application in an Airline MRO supply chain. *Int. J. Inf. Manag.* **2015**, *35*, 620–631. [CrossRef]
46. Sellitto, M.A.; Pereira, G.M.; Borchardt, M.; Da Silva, R.I.; Viegas, C.V. A SCOR-based model for supply chain performance measurement: Application in the footwear industry. *Int. J. Prod. Res.* **2015**, *53*, 4917–4926. [CrossRef]
47. Sutoni, A.; Subhan, A.; Setyawan, W.; Bhagyana, F.O. Mujiarto Performance Analysis Using the Supply Chain Operations Reference (SCOR) and AHP Method. *J. Phys. Conf. Ser.* **2021**, *1764*, 012155. [CrossRef]
48. Nguyen, T.T.H.; Bekrar, A.; Le, T.M.; Abed, M. Supply Chain Performance Measurement using SCOR Model: A Case Study of the Coffee Supply Chain in Vietnam. In Proceedings of the 1st International Conference On Cyber Management And Engineering (CyMaEn), Online, 26–28 May 2021; pp. 1–7.
49. Defrizal, D.; Hakim, L.; Kasimin, S. Analysis of Rice Supply Chain Performance Using the Supply Chain Operation Reference (Scor) Model and Analytical Hierarchy Process (Ahp) Method (Case Study: CV. Meutuah Baro Kuta Baro Aceh Besar District). *Int. J. Multicult. Multirelig. Underst.* **2020**, *7*, 222. [CrossRef]
50. Novar, M.F.; Ridwan, A.Y.; Santosa, B. SCOR and AHP Based Monitoring Dashboard to Measure Rice Sourcing Performance at Indonesian Bureau of Logistics. In Proceedings of the 12th International Conference on Telecommunication Systems, Services and Applications (TSSA), Kota Bandung, Indonesia, 4–5 October 2018; pp. 1–6.
51. Aliaga Rota, M.L.; Jané Portocarrero, J.J.; Merino Ascarrunz, R.C. Herramienta Para la Aplicación del Modelo SCOR en el Sector Confección del Perú. Master's Thesis, Pontifical Catholic University of Peru, Lima, Peru, 2020.
52. Kusrini, E.; Rifai, M.A.B.; Miranda, S. Performance measurement using supply chain operation reference (SCOR) model: A case study in a small-medium enterprise (SME) in Indonesia. *IOP Conf. Ser. Mater. Sci. Eng.* **2019**, *697*, 012014. [CrossRef]
53. Prasetya, W.; Natalia, C.; Dias, S.; Elnath, B.; Silalahi, A.; Monique, K. Performance measurement and analysis of coffee supply chain with SCOR method (case study of North Sumateria coffee). *IJRDO J. Bus. Manag.* **2017**, *3*, 11.
54. Castro Romero, N.A. Diagnóstico y Propuesta de Mejora en la Gestión de Inventarios y Distribución de Almacén en una Importadora de Juguetes Aplicando el Modelo SCOR y Herramientas de Pronósticos. Master's Thesis, Pontifical Catholic University of Peru, Lima, Peru, 2015.
55. Lama, J.L.; Esteban, F.C.L. Análisis del modelo SCOR para la Gestión de la Cadena de Suministro. In Proceedings of the IX Congreso de Ingeniería de Organización, Gijón, Spain, 8–9 September 2005; Volume 8, pp. 1–10.
56. Rojas López, M.M. Propuesta de Implementación del Modelo SCOR Para Incrementar la Efectividad de los Procesos de la Cadena de Suministro de la Empresa Import y Export Panita E.I.R.L. Industrial Engineering Thesis, Universidad Nacional de Trujillo, Trujillo, Peru, 2018.
57. Rotaru, K.; Wilkin, C.; Ceglowski, A. Analysis of SCOR's approach to supply chain risk management. *Int. J. Oper. Prod. Manag.* **2014**, *34*, 1246–1268. [CrossRef]
58. Girjatovics, A.; Pesoa, L.M.; Kuznecova, O. Establishing Supply Chain process framework based on SCOR model: Case study. In Proceedings of the 59th International Scientific Conference on Information Technology and Management Science of Riga Technical University (ITMS), Riga, Latvia, 10–12 October 2018; pp. 1–4.
59. Racine, N.; Hetherington, E.; McArthur, B.A.; McDonald, S.; Edwards, S.; Tough, S.; Madigan, S. Maternal depressive and anxiety symptoms before and during the COVID-19 pandemic in Canada: A longitudinal analysis. *Lancet Psychiatry* **2021**, *8*, 405–415. [CrossRef]
60. Yoon, J.A.; Choi, H.; Shin, Y.B.; Jeon, H. Development of a questionnaire to identify ocular torticollis. *Eur. J. Pediatr.* **2020**, *180*, 561–567. [CrossRef] [PubMed]
61. Hasibuan, A.; Arfah, M.; Parinduri, L.; Hernawati, T.; Suliawati; Harahap, B.; Sibuea, S.R.; Sulaiman, O.K.; Purwadi, A. Performance analysis of Supply Chain Management with Supply Chain Operation reference model. *J. Phys. Conf. Ser.* **2018**, *1007*, 012029. [CrossRef]
62. Waaly, A.N.; Ridwan, A.Y.; Akbar, M.D. Development of sustainable procurement monitoring system performance based on Supply Chain Reference Operation (SCOR) and Analytical Hierarchy Process (AHP) on leather tanning industry. *MATEC Web Conf.* **2018**, *204*, 01008. [CrossRef]
63. Leal, J.E. AHP-express: A simplified version of the analytical hierarchy process method. *MethodsX* **2019**, *7*, 100748. [CrossRef]
64. Kusrini, E.; Caneca, V.; Helia, V.; Miranda, S. Supply chain performance measurement using SCOR 12.0 Model A case study in a leather SME in Indonesia. *IOP Conf. Ser. Mater. Sci. Eng.* **2019**, *697*, 012023. [CrossRef]
65. Expoflores. Reporte Estadístico Anual 2020. Available online: https://expoflores.com/wp-content/uploads/2021/03/Anual-Expoflores.pdf (accessed on 7 February 2022).
66. Tapia Bustamente, X.P.T. Estudio de la Cadena Productiva y su Impacto en la Rentabilidad de las Empresas Florícolas en la Provincia de Cotopaxi. Master's Thesis, Universidad Técnica de Ambato, Ambato, Ecuador, 2019.
67. Smits, P. Large Colombian Companies on Take Over Spree in Ecuador. Floribusiness Edition 2—April 2021. 2021. Available online: https://www.hortipoint.nl/floribusiness/large-colombian-companies-on-takeover-spree-in-ecuador/ (accessed on 2 April 2021).

68. Superintendencia de Compañías del Ecuador. Información Financiera Empresas Ecuador 2021. Available online: https://www.supercias.gob.ec/portalscvs/ (accessed on 30 April 2021).
69. Jenkins, S.; Solomonides, T. Automating Questionnaire Design and Construction. *Int. J. Mark. Res.* **2000**, *42*, 1–13. [CrossRef]
70. Marques-Perez, I.; Guaita-Pradas, I.; Gallego, A.; Segura, B. Territorial planning for photovoltaic power plants using an outranking approach and GIS. *J. Clean. Prod.* **2020**, *257*, 120602. [CrossRef]

horticulturae

MDPI

Article

Efficacy of Different Concentrations of NAA on Selected Ornamental Woody Shrubs Cuttings

Endre Kentelky [1,*], Denisa Jucan [2,*], Maria Cantor [2] and Zsolt Szekely-Varga [1]

[1] Department of Horticulture, Faculty of Technical and Human Sciences, Sapientia Hungarian University of Transylvania, Sighișoarei 1C, 540485 Târgu Mureș, Romania; szekelyvarga.zsolt@gmail.com

[2] Department of Horticulture and Landscaping, Faculty of Horticulture, University of Agricultural Sciences and Veterinary Medicine of Cluj-Napoca, 400372 Cluj-Napoca, Romania; maria.cantor@usamvcluj.ro

* Correspondence: kentelky@ms.sapientia.ro (E.K.); denisa.jucan@usamvcluj.ro (D.J.)

Abstract: Ornamental woody shrubs are used in landscape design worldwide. Their propagation can be made generatively and vegetatively. Vegetative propagation methods are mostly used by nurseries, as such methods are quick and the newly propagated plants inherit the genetics of the mother plant. However, rooting in some woody plants is slow and, unfortunately, sometimes produces only a small number of rooted cuttings. In this study, shoot cuttings from six selected ornamental woody shrubs were subjected to different concentrations of rooting stimulators (0.5 (NAA5) and 0.8 (NAA8) % concentrations of 1-Naphthylacetic acid; cuttings without treatment were considered as control) and propagated in two different periods (spring and summer). Our results show that significant changes were obtained in the plants under the different treatments. Most of the plants showed a positive response to both treatments, expect for *Cotinus coggygria* 'Royal Purple', which, compared to control, registered decreases in all the tested parameters under NAA5 treatment. *Ilex aquifolium* was the species that showed increments in all the parameters when NAA treatments were applied. In conclusion, our research suggests that NAA increases rooting in ornamental woody shrubs, although in some cases rooting could be a species-dependent process.

Keywords: 1-Naphthylacetic acid; stimulants; propagation; rooting; shrubs

Citation: Kentelky, E.; Jucan, D.; Cantor, M.; Szekely-Varga, Z. Efficacy of Different Concentrations of NAA on Selected Ornamental Woody Shrubs Cuttings. *Horticulturae* **2021**, *7*, 464. https://doi.org/10.3390/horticulturae7110464

Academic Editor: Piotr Salachna

Received: 14 October 2021
Accepted: 1 November 2021
Published: 4 November 2021

1. Introduction

Interest in ornamental woody plants has been increasing in recent years and they are an important part of the horticulture industry. Ornamental shrubs are valued for their countless landscape uses and need to be part of our modern managed landscapes, for instance, as roadside trees in public parks which provide shade, shelter, clean pollutants in the air, and are a source of beauty [1,2].

With growing demand for ornamental shrubs, nurseries and horticulturists need new propagation methods in order to meet it. These types of plants can be propagated generatively and also vegetatively [3,4]. All woody plants are capable of producing flowers and seeds; however, they require favorable environmental conditions and take many years to develop [2]. Most of them are propagated by vegetative methods, by cuttings, because such methods are quicker and also because the plants will retain the characteristics and genetics of the mother plants [5,6].

The rooting of ornamental woody plants can sometimes be a hard and slow method, and does not have a high success rate. Propagation by cuttings is a vegetative method widely used for different plant species. Ornamental woody plant nurseries have developed different techniques to successfully improve the rooting of cuttings. However, in spite of controlled environmental conditions, high economic losses are still being sustained as a result of insufficient root formation [7,8]. In addition to environmental factors, the successful rooting of woody plant cuttings could be affected by different elements, such as nutritional levels of the mother plant, cutting type, rooting medium, and even by the manipulation and treatments applied [9].

Hormones could improve the percentage of radicals and also reduce propagation time [10]. Plant hormones are substances naturally produced by plants which control plant functions and development, such as root growth, fruit maturation, and plant growth [11,12]. Hormones are important and crucial elements which are required to control plant development through the life cycle, from embryogenesis to reproductive development [13–15].

Adventitious root formation is a physiological process enabling the propagation of cuttings of many plant species [7]. Previous reports suggested that adventitious root formation in woody plants could be associated with the action of endogenous auxin and can be triggered by the application of exogenous auxin, such as 1–Naphthylacetic acid (NAA) [16–18]. NAA is used to influence/induce and to ensure a greater rooting capacity of cuttings and the better establishment of many shrubs and trees [8,10,19]. NAA could even effectively improve the survival rate of cuttings or shorten the rooting period [20].

Syringa vulgaris L., commonly known as Lilac, is a deciduous shrub including more than 40 species distributed around Europe and Asia [21,22]. *Ilex aquifolium*, native to southern Europe, northwest Africa, and southwest Asia, commonly known as English holly, is a dioecious plant species with persistent leaves and female and male flowers on different plants [23,24]. *Cotinus coggygria* (Smoketree) is a woody shrub growing wildly in Europe and Asia [25,26].

The aim of the present study was to test the effect of NAA in two different concentrations on six ornamental woody shrubs often used in Romanian landscape design. *Syringa vulgaris* 'Mme Lemoine', *Syringa vulgaris* 'President Grevy', *Ilex aquifolium*, *Cotinus coggygria*, *Cotinus coggygria* 'Kanari', and *Cotinus coggygria* 'Royal Purple' were analyzed in the experiment. The influence of NAA on rooting percentage, root volume, number of roots, root length, and rooted cutting diameter were investigated. We aimed to determine the concentration most suitable for the vegetative propagation of woody ornamental plants.

2. Materials and Methods

2.1. Experimental Site and Plant Material

The study was conducted between May and October 2019 in the experimental greenhouse belonging to Sapientia Hungarian University of Transylvania, Târgu Mureș (46°31′17″ N 24°35′54″ E). The cuttings were obtained from a local nursery (Biota, Găiești village, Romania). The cuttings were immediately transported to the experimental sites to prevent desiccation. As plant material, the following ornamental woody shrubs were selected:

- *Syringa vulgaris* 'Mme Lemoine' (SVM): double white flowers, light green heart-shaped leaves, grows up to 2.5–3 m and 3 m wide.
- *Syringa vulgaris* 'President Grevy' (SVP): double lavender-blue flowers, light green heart-shaped leaves, grows up to 3–3.5 m and 2.5 m wide.
- *Ilex aquifolium* (IA): cluster white flowers, produces red fruits, glossy green prickly leaves, grows up to 10–25 m and 2 m wide.
- *Cotinus coggygria* (CC): smoky pink flowers, green or reddish-purple leaves, grows up to 3–5 m and 4 m wide.
- *Cotinus coggygria* 'Kanari' (CCK): white flowers, green leaves, grows up to 2.5–4 m and 2.5–4 m wide.
- *Cotinus coggygria* 'Royal Purple' (CCR): feathery pink flowers, wine purple leaves, grows up to 3–5 m and 4.5–6 m wide.

2.2. Experimental Design and Rooting Conditions

The first experiment started on 21 May (spring propagation), and in the summer on 7 July the experiment was repeated, with the same ornamental shrub species and rooting stimulants.

For each species and cultivar 10 sub-apical shoots (herbaceous spring and semi-hardwood summer cuttings) per replication, with three replications, were used—a total of

540 cuttings. Disease and pest-free propagation material was between 8–10 cm in length and was collected with a secateur from the nursery. The leaves on the lower one-third to one-half of the stems were removed. After treatments were applied, the cuttings were planted in 60 × 40 cm plastic trays, filled with perlite rooting medium. Planting distance was 2 cm between the cuttings. We had filled the plastic tray with perlite to a depth of 20 cm (granulation: 1–3 mm, density: 0.05 kg/L, and pH: 7–7.5) and this was well irrigated before planting the cuttings. No artificial lights were installed. Propagation trays were placed in the greenhouse with an automatic humidifier controller in order to provide the 80–90% humidity required for rooting. Humidity and temperature were measured using a Testo 175H1 (Testo Romania, Cluj-Napoca, Romania); the average temperature was between 22–28 °C.

From each species, 30 cuttings per treatment were immersed in Incit–5 (AMVAC Chemical UK Ltd., Surrey, UK) and Incit–8 (AMVAC Chemical UK Ltd., Surrey, UK) rooting hormones, approximately up to 1–1.5 cm. Incit-5 composition was 0.5% 1-Naphthaleneacetic acid (NAA5) and Incit-8 0.8% of 1–Naphthaleneacetic acid (NAA8), both recommended as rooting stimulants for ornamental woody plants. Cuttings without treatment were considered as control.

2.3. Data Evaluation

Data for the rooted cuttings propagated in spring were reported on 13th September (116 days after preparing the cuttings) and for the summer rooted cuttings on 28th October (114 days after preparing the cuttings).

Rooting percentage (the percentage of cuttings that developed at least one root), root volume (cm^3—a measuring cylinder was filled with water, the plant was submerged in it and under the pressure of the cutting water, filled out), number of roots, root length (cm) and rooted cutting diameter (cm) were determined. Root length was measured with a tape measure and cutting diameter with a digital caliper (GartenVIP DiyLine, Alba Iulia, Romania).

2.4. Statistical Analysis

The data were tested for normality of errors and homogeneity of variance. As all data were normally distributed, ANOVA followed by Tukey's test was used to compare variances. The significance of the differences between the treatments was tested by applying two-way ANOVA, at a confidence level of 95%. When the ANOVA null hypothesis was rejected, Tukey's post hoc test was carried out to establish the statistically significant differences at $p < 0.05$.

3. Results

3.1. Rooting Percentage of Cuttings

Concerning rooting percentage, hormone type influenced the process in different ways (Figure 1). However, it no significant differences were recorded between the spring and summer cuttings propagation. SVM (Figure 1a) reported small increases compared to control. In the case of SVP (Figure 1b), differences were determined when comparing the two treatments to control, although 0.5% 1-Naphthylacetic acid, compared with the other treatment (0.8% 1-Naphthylacetic acid), highly increased the rooting percentage, with 90% of summer cuttings rooting. Similar data were reported for the IA cuttings (Figure 1c), where at both propagation times the greatest percentage of rooting was observed in plants subjected to NAA5 treatment. Regarding *Cotinus coggygria* (CC), significant increases were reported for NAA8, almost double the rooting percentage compared to control (Figure 1d). In contrast, the data reported for NAA5 were similar to those for the untreated CC (Figure 1d). In the case of CCK (Figure 1e), significant increases were determined just with NAA8. Nevertheless, for *Cotinus coggygria* 'Royal Purple' (CCR), no significance was observed between the control and the NAA8 treated plants, yet rooting percentage decreased at the CCR subjected to 0.5% 1-Naphthylacetic acid treatment (Figure 1f).

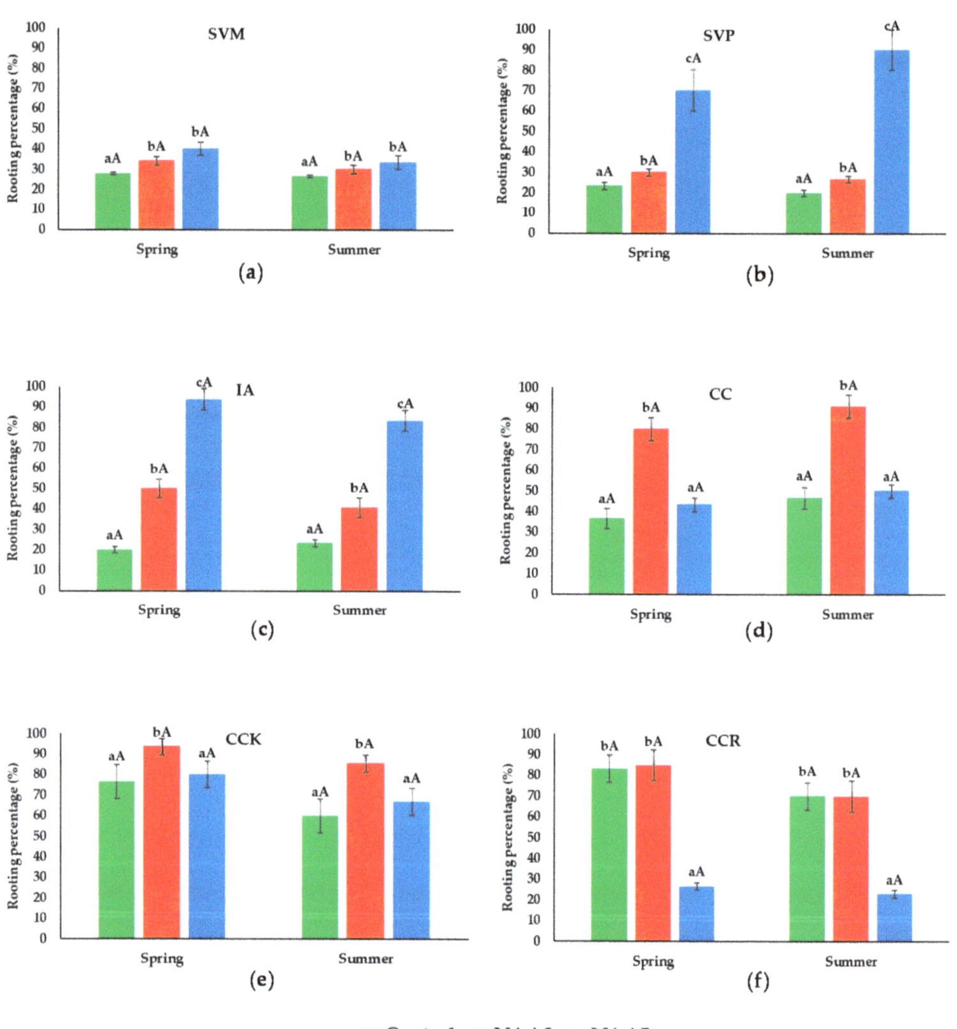

Figure 1. Effect of rooting stimulants (NAA8—0.8% concentration of 1-Naphthylacetic acid and NAA5—0.5% concentration of 1-Naphthylacetic acid) on rooting percentage for the six selected ornamental shrubs: (**a**) *Syringa vulgaris* 'Mme Lemoine' (SVM); (**b**) *Syringa vulgaris* 'President Grevy' (SVP); (**c**) *Ilex aquifolium* (IA); (**d**) *Cotinus coggygria* (CC); (**e**) *Cotinus coggygria* 'Kanari' (CCK); (**f**) *Cotinus coggygria* 'Royal Purple' (CCR). Bars represent the means ± SE (n = 30). Different lowercase letters above the bars indicate significant differences between the treatments, and different uppercase letters indicate the significant differences between the spring and summer propagated cuttings, according to Tukey's test (α = 0.05).

3.2. Root Volume

Under our experimental conditions, no significant differences were observed in root volume when comparing the two propagation periods (Figure 2). The root volume of *Syringa vulgaris* 'Mme Lemoine' (Figure 2a) increased under the treatments. In the case of cuttings propagated in the spring, significant differences were observed with NAA5, and summer cuttings showed increased root volume with both treatments; however, the NAA5 recorded higher increases. Additionally, increased root volume was observed with SVP (Figure 2b) during the treatments, yet the largest increase was observed in the summer cuttings subjected to NAA5, where the volume of the roots was approximately 16 times higher than in the untreated plants' root systems. Considering *Ilex aquifolium* (Figure 2c), it can be determined that treatments greatly increased root volume. IA spring cuttings under the 0.5% NAA treatment reported root volumes of 1.99 cm^3 compared to control, in which case the root volume was just 0.1%. Significant differences were also observed in the treated plants (CC) compared to the controls (Figure 2d). Rooting hormone NAA8 greatly increased the root volume of CCK (Figure 2e). In the case of *Cotinus coggygria* 'Royal Purple' (CCR), the development of volume of the roots was inhibited by NAA5 treatment (Figure 2f), compared to control.

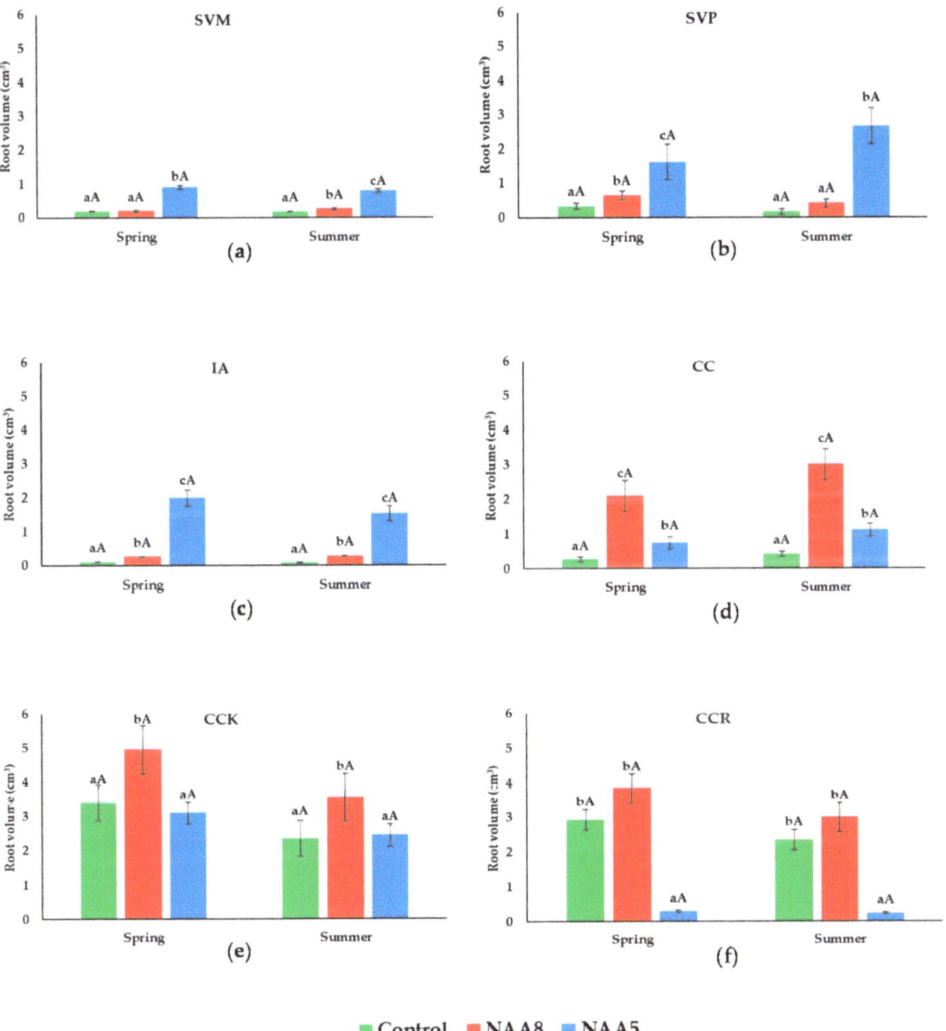

Figure 2. Effect of rooting stimulants (NAA8 and NAA5) on root volume in the six selected ornamental shrubs: (**a**) *Syringa vulgaris* 'Mme Lemoine' (SVM); (**b**) *Syringa vulgaris* 'President Grevy' (SVP); (**c**) *Ilex aquifolium* (IA); (**d**) *Cotinus coggygria* (CC); (**e**) *Cotinus coggygria* 'Kanari' (CCK); (**f**) *Cotinus coggygria* 'Royal Purple' (CCR). Bars represent the means ± SE (n = 30). Different lowercase letters above the bars indicate significant differences between the treatments, and different uppercase letters indicate the significant differences between the spring and summer propagated cuttings, according to Tukey's test (α = 0.05).

3.3. *Number of Roots*

As expected, the number of roots was significantly affected by hormone products. In the case of SVM (Figure 3a) treated with NAA5, root numbers were about four times higher than in the controls. Increments in number roots were also observed in SVP (Figure 3b). However, in the spring cuttings, only in plants treated with 0.5% 1-Naphthylacetic acid were increases reported. On the other hand, by the summer, both treatments influenced the number of roots in SVP plants in a positive way. Significant differences between treated and untreated IA plants were observed (Figure 3c) with NAA5 treatment, which increased root number at both spring and summer. Under our experimental conditions, increases in the number of roots of CC were observed with NAA8 treatment in both propagation periods, and also in the summer cycle with 0.5% 1-Naphthylacetic acid treatment (Figure 3d). Spring cuttings of CCK (Figure 3e) showed root number increases when subjected to NAA8 treatment, though no significant results were recorded for summer propagation. In the case of CCR (Figure 3f), it was concluded that NAA8 has no influence on the number of roots for spring or summer cuttings. Moreover, NAA5 had inhibited the development of roots compared to control. It is important to mention that no significant differences were determined between spring and summer propagations (Figure 3).

3.4. *Root Length*

From the results for the cuttings, no significant changes were observed in root length between the spring and summer periods of treatment (Figure 4). However, for SVM, increases were reported compared to control. The influence of NAA5 was greater than the other hormone type (Figure 4a). For SVP (Figure 4b), no effect was measured at the spring cutting under NNA8 treatment; by contrast, increases were reported for the spring cutting treated with NAA5. With the summer cuttings subjected to both rooting stimulants, significant results were observed. IA cuttings registered increases of root length under both treatments (Figure 4c). In the case of CC, high increments in relation to controls were reported in both propagation periods, with 0.8% 1-Naphthylacetic acid and with 0.5% 1-Naphthylacetic acid in the summer cuttings (Figure 4d). *Cotinus coggygria* 'Kanari' (Figure 4e) showed increases in spring and summer cuttings under the NAA8 treatment. NAA5 treatment had a significant negative influence on the root length of *Cotinus coggygria* 'Royal Purple'. In contrast, no effect was observed in cuttings treated with NAA8 compared to control (Figure 4f).

3.5. *Diameter of Cuttings*

As expected, no differences were shown in the diameter of rooted cuttings when comparing spring and summer propagation (Figure 5), and no effect of rooting stimulants was reported for SVM (Figure 5a). On the contrary, SVP cutting diameters were highly influenced by NAA5 treatment in both propagation experiments (spring, 3.14 cm; summer, 3.82 cm), and increases were also observed in summer cuttings subjected to NAA8 (Figure 5b). Comparing the control *Ilex aquifolium* to the treated cuttings, it could be concluded that the diameter of spring and summer cuttings reported high increases with both treatments (Figure 5c). In the case of CC (Figure 5d), cutting diameters showed significant increases under the NAA8 treatment. No differences were observed in the CCK rooted cuttings diameter (Figure 5e). NAA8 increased the diameter of cuttings; on the other hand, NAA5 inhibited the thickness of the cuttings' diameters for CCR (Figure 5f).

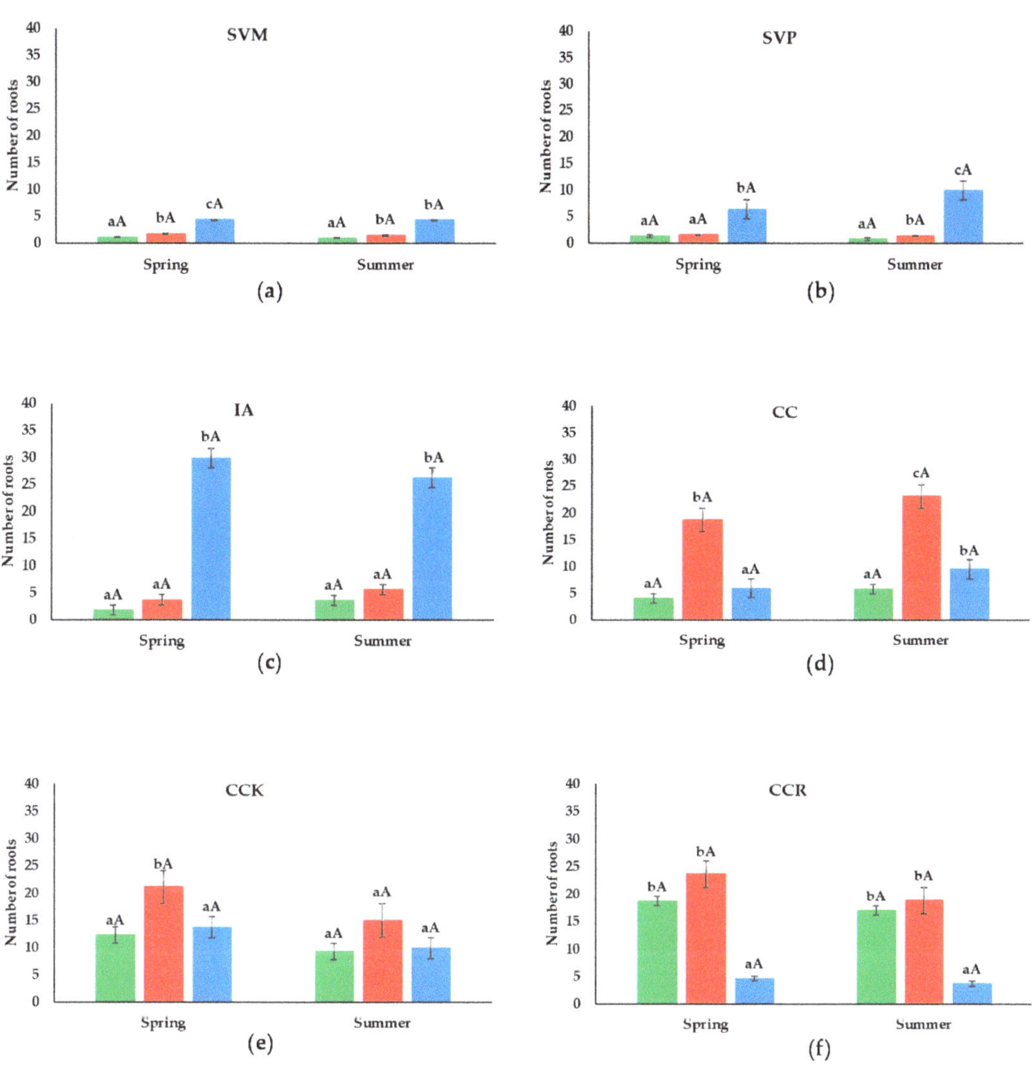

Figure 3. Effect of rooting stimulants (NAA8 and NAA5) on the number of roots in the six selected ornamental shrubs: (**a**) *Syringa vulgaris* 'Mme Lemoine' (SVM); (**b**) *Syringa vulgaris* 'President Grevy' (SVP); (**c**) *Ilex aquifolium* (IA); (**d**) *Cotinus coggygria* (CC); (**e**) *Cotinus coggygria* 'Kanari' (CCK); (**f**) *Cotinus coggygria* 'Royal Purple' (CCR). Bars represent the means ± SE (n = 30). Different lowercase letters above the bars indicate significant differences between the treatments, and different uppercase letters indicate the significant differences between the spring and summer propagated cuttings, according to Tukey's test (α = 0.05).

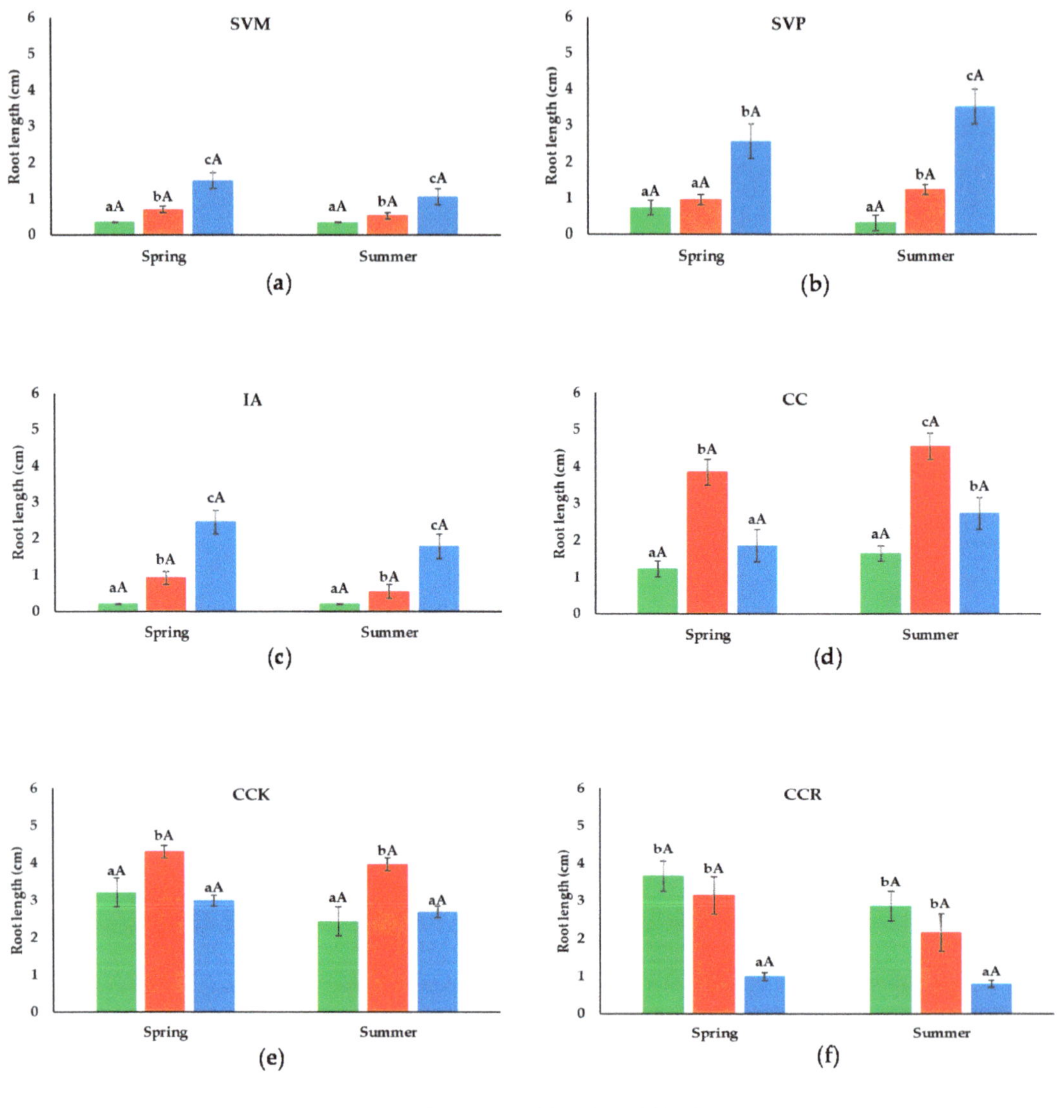

Figure 4. Effect of rooting stimulants (NAA8 and NAA5) on root length in the six selected ornamental shrubs: (**a**) *Syringa vulgaris* 'Mme Lemoine' (SVM); (**b**) *Syringa vulgaris* 'President Grevy' (SVP); (**c**) *Ilex aquifolium* (IA); (**d**) *Cotinus coggygria* (CC); (**e**) *Cotinus coggygria* 'Kanari' (CCK); (**f**) *Cotinus coggygria* 'Royal Purple' (CCR). Bars represent the means ± SE (n = 30). Different lowercase letters above the bars indicate significant differences between the treatments, and different uppercase letters indicate the significant differences between the spring and summer propagated cuttings, according to Tukey's test (α = 0.05).

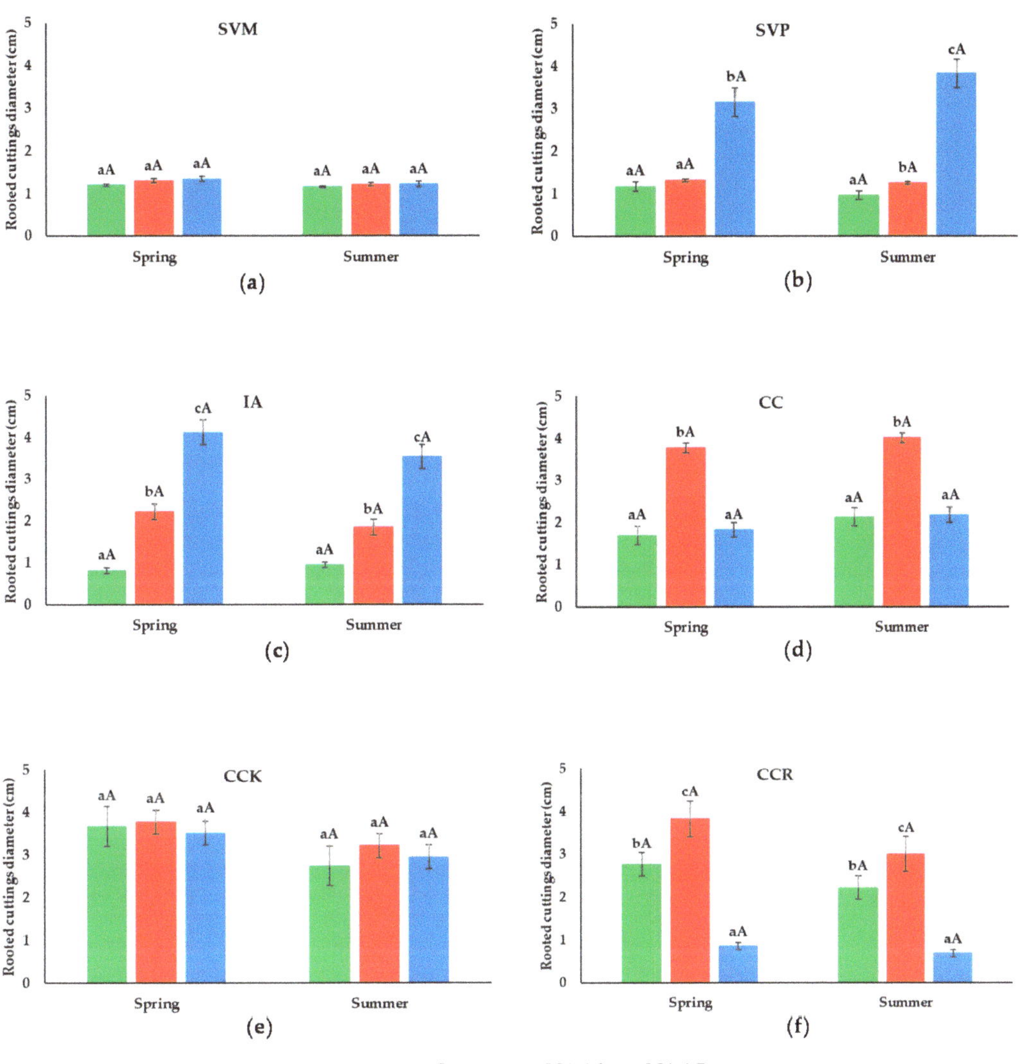

Figure 5. Effect of rooting stimulants (NAA8 and NAA5) on the diameter of rooted cuttings in the six selected ornamental shrubs: (**a**) *Syringa vulgaris* 'Mme Lemoine' (SVM); (**b**) *Syringa vulgaris* 'President Grevy' (SVP); (**c**) *Ilex aquifolium* (IA); (**d**) *Cotinus coggygria* (CC); (**e**) *Cotinus coggygria* 'Kanari' (CCK); (**f**) *Cotinus coggygria* 'Royal Purple' (CCR). Bars represent the means ± SE (n = 30). Different lowercase letters above the bars indicate significant differences between the treatments, and different uppercase letters indicate the significant differences between the spring and summer propagated cuttings, according to Tukey's test (α = 0.05).

4. Discussion

Rooting stimulants can be used to increase the rooting capacity of different plants [27–31] and for obtaining the maximum number of rooted cuttings in a short period of time [32]. However, some studies concluded that rooting media is also an important factor which could affect rooting percentage in different ornamental cuttings [1,33,34]. Rooting hormones could positively influence the rooting process of woody plants, but in some cases, this depends on species or their natural rooting ability [35,36]. The rooting percentage of *Parthenocissus quinquefolia* was increased with the use of two different stimulants, although it was mentioned in the study that growing media combined with the different stimulants could also have a positive effect on the rooting percentage of cuttings [37]. Nevertheless, blue light combined with NAA treatments could significantly improve the rooting and leaf-bud of *Chrysanthemum* cuttings [38]. Thus, propagation period is critical for the rooting process. Our observations indicated that it has no effect on the rooting capacity of the selected ornamental woody shrubs. Though it has been concluded in some studies that the season could affect the rooting of the cuttings [39–41], this could also be a species-dependent factor.

From our results, it can be concluded that rooting percentage increased in the woody shrubs subjected to treatments; however, an inhibition was observed in the case of CCR treated with NAA, compared to control. Of course, it is important to mention that even if NAA5 and NAA8 boosted rooting percentage, not all plants behaved in the same way. Adventitious root formation is a critical phase for the survival and growth of the propagated cuttings [42], involving morphological, physiological, and biological changes [43,44]. The application of NAA improved the rapid recovery of the wounded surface and also affected the rapid appearance of adventitious roots, which guaranteed the cuttings' survival rates. It was reported in a study that 0.3% of NAA concentration resulted in the highest rooting percentage of *Jasminum parkeri* [31]. It was also reported that just 0.01% NAA in combination with 0.01% GA_3 can improve rooting percentage of *Hydrangea* [45]. NAA used in micropropagation improved in vitro root induction in *Magnolia sirindhorniae* [46]. Nevertheless, for *Ficus benjamina* L., it was reported that the highest rooting percentage was obtained with just 0.001% of NAA [47], which, compared to our concertation, is very low, yet still increased the rooting of the cuttings.

The data obtained show that rooting stimulants can clearly have a positive effect on the root volume of ornamental woody shrub cuttings. Under our experimental conditions, NAA5 greatly increased root volume in both *Syringa vulgaris* and *Ilex aquifolium*. On the other hand, NAA8 reported higher increases for CC, CCK, and CCR, but with *Cotinus coggygria* 'Kanari' and 'Royal Purple', root volume was inhibited. Previous studies have also reported increases in root volume under different stimulants [48,49]. In one study it was determined that a combination of IBA 1500 ppm + NAA 1000 ppm resulted in the highest root volume in *Piper nigrum* L. [50].

Application of rooting stimulants (1-Naphthyl and 2-Naphthhyl) slightly increase the number of roots of apple and mung bean [51]. In a study by Trofimuk et al. [52] it was reported that rooting stimulants influence in a positive way the number and length of roots of *Abies gracilis*, which is useful in accelerating the production of plant material, and reduces rooting time in woody ornamental plants. Previous research found that NAA could also boost the number of adventitious roots, even the growth and development of micropropagated plants [53–55]. Under our experimental conditions, root length increased under the NAA5 treatment in SVM, SVP, IA, and CC. However, with CCK and CCR, root length was similar or was even inhibited under the same treatment. Another study conducted reveled that indolebutyric acid increased the number of roots and root length of blueberry cuttings, but on the contrary, no effect was observed in rooting percentage and survival percentage of the cuttings [56]. Planting dates and NAA treatments could significantly improve root number and length [57], yet under our experimental conditions, it was determined that planting time did not influence root number or root length. Our data are similar to previous studies which have concluded that number and length of roots are positively affected by the application of NAA, at least at 0.5% concentrations [38,58–61].

From the data obtained in our experiment, it could be concluded that NAA influenced every woody shrub cutting diameter in a different way. Furthermore, no significant differences were observed in SVM. NAA5 clearly affected cutting diameters in SVP in both propagation times. NAA8 summer cuttings also showed differences compared to control, but not as high as those treated with NAA5. Both concentrations of 1-Naphthylacetic acid greatly increased the diameters of IA rooted cuttings. In the case of *Cotinus coggygria*, significant increases were determined with NAA8. On the other hand, no significant difference was observed for CCK. Significant differences were observed in CRR cutting treated with NAA8; however, NAA5 inhibited the treated shrubs. In some studies, it was reported that diameter could have an effect on root number and on the length of cuttings [62,63]. This was clearly observed in our experiment—that where cutting diameter increased, root length and number increased with it. Similar data was recorded for *Punica granatum* L., where with increases in diameter, the length and number of roots increased under IBA + NAA treatment [64]. The rooting process depends on the cutting's diameter and on the nutrients which sustain the biological processes involved in adventitious root formation [65–67]. Altogether, it can be concluded that rooting stimulants could have beneficial effects on the development, growth and survival percentage of the plants studied [68–70].

5. Conclusions

Ornamental woody plant nurseries strive to produce rooted cuttings in a short time and to ensure that they are of good quality. The present study provides new experimental data on the comparison of two rotting stimulants on six woody shrubs often used in landscape design. According to the results, it can be concluded that 1-Naphthylacetic acid used in different concentrations could have a positive effect on the rooting of the plants selected in this experiment. Results show that NAA8 treatment positively affected the root percentage of CC and CCK, and NAA5 influenced negatively root percentage in CCR in both propagation periods. However, root percentage in CC and CCK was not significantly influenced in either treatment period. On the other hand, significant (negative) changes were reported in CCR rooting percentage for cuttings under NAA8 treatment. These data show that rhizogenesis could be a species- or cultivar-dependent process. Regarding the number of roots, NAA5 showed better results for *Syringa* tested cultivars and for *Ilex aquifolium*, while NAA8 had a greater influence on the analyzed species and cultivars of *Cotinus*. Root length increased when SVM, SVP and IA cuttings were treated with NAA5, while NAA8 increased the length of the root systems in CC and CCK. On the basis of the results presented here, it could be stated that rooting hormones/stimulants strengthen the possibility of achieving a quicker vegetative propagation method, but future experiments need to be conducted.

Author Contributions: Conceptualization, E.K. and Z.S.-V.; methodology, E.K. and M.C.; formal analysis, M.C. and Z.S.-V.; resources, E.K. and D.J.; writing—original draft preparation, Z.S.-V.; writing—review and editing, Z.S.-V. and D.J. All authors have read and agreed to the published version of the manuscript.

Funding: This research was partially funded by the University of Agricultural Sciences and Veterinary Medicine of Cluj-Napoca.

Institutional Review Board Statement: Not applicable.

Informed Consent Statement: Not applicable.

Data Availability Statement: Not applicable.

Acknowledgments: This work was supported by the Institute of Advance Horticulture Research of Transylvania, University of Agricultural Science and Veterinary Medicine of Cluj–Napoca and the Sapientia Hungarian University of Transylvania.

Conflicts of Interest: The authors declare no conflict of interest.

References

1. Naik, E.K. Success rate of different ornamental cuttings based on different growing media. *J. Pharmacogn. Phytochem.* **2018**, *7*, 2479–2482.
2. Swaroop, K. *Unit-1 Woody Ornamentals Trees*; Indira Gandhi National Open University: New Delhi, India, 2021.
3. Ruchala, S.L. *Propagation of Several Native Ornamental Plants*; Master's Thesis, The University of Maine, Orono, ME, USA, 2002; p. 448.

4. Mateescu, R. *Ornamental Tree and Shrubs*; M.A.S.T. Publishing: Bucharest, Romania, 2002; pp. 28–64.
5. Brown, C.L.; Sommer, H.E. Vegetative propagation of dicotyledonous trees. In *Tissue Culture in Forestry*; Springer: Dordrecht, The Netherlands, 1982; pp. 109–149. [CrossRef]
6. Stuepp, C.A.; Wendling, I.; Xavier, A.; Zuffellato-Ribas, K.C. Vegetative propagation and application of clonal forestry in Brazilian native tree species. *Pesqui. Agropecu. Bras.* **2018**, *53*, 985–1002. [CrossRef]
7. Husen, A.; Iqbal, M.; Siddiqui, S.N.; Sohrab, S.S.; Masresha, G. Effect of indole-3-butyric acid on clonal propagation of mulberry (Morus alba L.) stem cuttings: Rooting and associated biochemical changes. In *Proceedings of the National Academy of Sciences, India Section B: Biological Sciences*; Springer: Dordrecht, The Netherlands, 2017; Volume 87, pp. 161–166. [CrossRef]
8. Kaviani, B.; Negahdar, N. Propagation, micropropagation and cryopreservation of *Buxus hyrcana* Pojark, an endangered ornamental shrub. *S. Afr. J. Bot.* **2017**, *111*, 326–335. [CrossRef]
9. Kumar, K.V.; Fatmi, U. Effects of IBA and NAA on shoot growth of cuttings of various ornamental plants in water as rooting medium. *J. Pharmacogn. Phytochem.* **2021**, *10*, 685–687.
10. Kaushik, S.; Shukla, N. A review on effect of IBA and NAA and their combination on the rooting of stem cuttings of different ornamental crops. *J. Pharmacogn. Phytochem.* **2020**, *9*, 1881–1885.
11. Mirihagalla, M.K.P.N.; Fernando, K.M.C. Effect of *Aloe vera* gel for inducing rooting of stem cuttings and air layering of plants. *J. Dry Zone Agric.* **2020**, *6*, 13–26.
12. Anfang, M.; Shani, E. Transport mechanisms of plant hormones. *Curr. Opin. Plant Biol.* **2021**, *63*, 102055. [CrossRef] [PubMed]
13. Sajjad, Y.; Jaskani, M.J.; Asif, M.; Qasim, M. Application of plant growth regulators in ornamental plants: A review. *Pak. J. Agric. Sci.* **2017**, *54*, 327–333. [CrossRef]
14. Toungos, M.D. Plant growth substances in crop production: A Review. *IJIABR* **2018**, *6*, 1–8.
15. Mendel, P.; Schiavo-Capri, E.; Lalge, A.B.; Vyhnanek, T.; Kalousek, P.; Trojan, V.; Havel, L.; Filippi, A.; Braidot, E. Evaluation of selected characteristics in industrial hemp after phytohormonal treatment. *Pak. J. Agric. Sci.* **2020**, *57*, 1–7.
16. Pop, T.I.; Pamfil, D.; Bellini, C. Auxin control in the formation of adventitious roots. *Not. Bot. Horti. Agrobot. Cluj-Napoca* **2011**, *39*, 307–316. [CrossRef]
17. Pacurar, D.I.; Perrone, I.; Bellini, C. Auxin is a central player in the hormone cross-talks that control adventitious rooting. *Physiol. Plant.* **2014**, *151*, 83–96. [CrossRef]
18. Park, S.H.; Elhiti, M.; Wang, H.; Xu, A.; Brown, D.; Wang, A. Adventitious root formation of in vitro peach shoots is regulated by auxin and ethylene. *Sci. Hortic.* **2017**, *226*, 250–260. [CrossRef]
19. OuYang, F.; Wang, J.; Li, Y. Effects of cutting size and exogenous hormone treatment on rooting of shoot cuttings in Norway spruce [*Picea abies* (L.) Karst.]. *N. For.* **2015**, *46*, 91–105. [CrossRef]
20. Wei, Y.; Zhou, D.; Han, Y.; Wu, Z.; Zhang, Z. Effects of Naphthylacetic acid on rootage of cutting twigs of the *Hydrangea macrophylla*. *J. Zhongkai Univ. Agric. Eng.* **2016**, *29*, 27–29.
21. Su, G.; Cao, Y.; Li, C.; Yu, X.; Gao, X.; Tu, P.; Chai, X. Phytochemical and pharmacological progress on the genus *Syringa*. *Chem. Cent. J.* **2015**, *9*, 1–12. [CrossRef] [PubMed]
22. Zhu, W.; Wang, Z.; Sun, Y.; Yang, B.; Wang, Q.; Kuang, H. Traditional uses, phytochemistry and pharmacology of genus *Syringa*: A comprehensive review. *J. Ethnopharmacol.* **2021**, *266*, 113465. [CrossRef] [PubMed]
23. Peterken, G.F.; Lloyd, P.S. *Ilex aquifolium* L. *J. Ecology.* **1967**, *55*, 841–858. [CrossRef]
24. Tsaktsira, M.; Chavale, E.; Kostas, S.; Pipinis, E.; Tsoulpha, P.; Hatzilazarou, S.; Ziogou, F.T.; Nianiou-Obeidat, I.; Iliev, I.; Economou, A.; et al. Vegetative Propagation and ISSR-Based Genetic Identification of Genotypes *of Ilex aquifolium* 'Agrifoglio Commune'. *Sustainability* **2021**, *13*, 10345. [CrossRef]
25. Matić, S.; Stanić, S.; Mihailović, M.; Bogojević, D. *Cotinus coggygria* Scop.: An overview of its chemical constituents, pharmacological and toxicological potential. *Saudi J. Biol. Sci.* **2016**, *23*, 452–461. [CrossRef]
26. Antal, D.S.; Ardelean, F.; Jijie, R.; Pinzaru, I.; Soica, C.; Dehelean, C. Integrating Ethnobotany, Phytochemistry, and Pharmacology of *Cotinus coggygria* and Toxicodendron vernicifluum: What Predictions can be Made for the European Smoketree? *Front. Pharmacol.* **2021**, *12*, 746. [CrossRef]
27. Da Silva, J.A.T.; Pacholczak, A.; Ilczuk, A. Smoke tree (*Cotinus coggygria* Scop.) propagation and biotechnology: A mini-review. *S. Afr. J. Bot.* **2018**, *114*, 232–240. [CrossRef]
28. Negahdar, N.; Hashemabadi, D.; Kaviani, B. An improved procedure for in vivo and in vitro propagation of *Buxus hyrcana*, an ornamental shrub critically at risk of extinction. *BioTechnologia* **2019**, *100*, 417–428. [CrossRef]
29. Monder, M.J.; Pacholczak, A. Rhizogenesis and concentration of carbohydrates in cuttings harvested at different phenological stages of once-blooming rose shrubs and treated with rooting stimulants. *Biol. Agric. Hortic.* **2020**, *36*, 53–70. [CrossRef]
30. Hamidon, A.; Shah, R.M.; Razali, R.M.; Lob, S. Effect of different types and concentration of rooting hormones on *Momordica cochinensis* (Gac Fruit) Root Vine Cuttings. *Malays. Appl. Biol.* **2020**, *49*, 127–132.
31. Kashyap, U.; Chandel, A.; Sharma, D.; Bhardwaj, S.; Bhargava, B. Propagation of *Jasminum parkeri*: A Critically Endangered Wild Ornamental Woody Shrub from Western Himalaya. *Agronomy* **2021**, *11*, 331. [CrossRef]
32. Ignatova, G.A. Use of growth promoter activators for rooting decorative cultures. *Bull. Agrar. Sci.* **2018**, *3*, 43–47. [CrossRef]
33. Kentelky, E. The analysis of rooting and growth peculiarities of *Juniperus* species propagated by cuttings. *Bull. UASVM Hortic.* **2011**, *68*, 380–385. [CrossRef]

34. Kumar, S.; Malik, A.; Yadav, R.; Yadav, G. Role of different rooting media and auxins for rooting in floricultural crops: A review. *Int. J. Chem. Stud.* **2019**, *7*, 1778–1783.
35. Monder, M.J.; Pacholczak, A. Rhizogenesis and contents of polyphenolic acids in cuttings of old rose cultivars treated with rooting stimulants. *Acta Hortic.* **2017**, *1232*, 99–104. [CrossRef]
36. Zamir, R.; Rab, A.; Sajid, M.; Khattak, G.S.S.; Khalil, S.A.; Shah, S.T. Effect of Different Auxins on Rooting of Semi Hard and Soft Wood Cuttings of Guava (*Psidium guajava* L.) CV. *Safeda. Nucleus* **2017**, *54*, 46–51.
37. Jeberean, M.G.; Bala, M.; Berar, C.; Tota, C.E.; Silivasan, M. Research on the influence of rooting stimulants at *Parthenocissus quinquefolia* in different crop conditions. *Int. Multidiscip. Sci. GeoConference* **2017**, *7*, 915–919. [CrossRef]
38. Gil, C.S.; Jung, H.Y.; Lee, C.; Eom, S.H. Blue light and NAA treatment significantly improve rooting on single leaf-bud cutting of *Chrysanthemum* via upregulated rooting-related genes. *Sci. Hortic.* **2020**, *274*, 109650. [CrossRef]
39. Tsaktsira, M.; Alevropoulos, A.; Tsoulpha, P.; Scaltsoyiannes, V.; Scaltsoyiannes, A.; Iliev, I. Inter-and intra-genetic variation on rooting ability of *Ilex aquifolium* L. Varieties and cultivars. *Propag. Ornam. Plants* **2018**, *18*, 131–138.
40. Swarts, A.; Matsiliza-Mlathi, B.; Kleynhans, R. Rooting and survival of *Lobostemon fruticosus* (L) H. Buek stem cuttings as affected by season, media and cutting position. *S. Afr. J. Bot.* **2018**, *119*, 80–85. [CrossRef]
41. Kumar, P.; Mishra, J.P.; Sonkar, M.K.; Mishra, Y.; Shirin, F. Relationship of Season and Cuttings' Diameter with Rooting Ability of Culm-Branch Cuttings in *Bambusa tulda* and *Bambusa nutans*. *J. Plant Growth Regul.* **2021**, 1–8. [CrossRef]
42. Koroch, A.; Juliani, H.R.; Kapteyn, J.; Simon, J.E. In vitro regeneration of *Echinacea purpurea* from leaf explants. *PCTOC* **2002**, *69*, 79–83. [CrossRef]
43. Geiss, G.; Gutierrez, L.; Bellini, C. Adventitious root formation: New insights and perspectives. *Annu. Plant Rev.* **2009**, *37*, 127–156. [CrossRef]
44. Zhang, W.; Fan, J.; Tan, Q.; Zhao, M.; Zhou, T.; Cao, F. The effects of exogenous hormones on rooting process and the activities of key enzymes of *Malus hupehensis* stem cuttings. *PLoS ONE* **2017**, *12*, e0172320. [CrossRef]
45. Punetha, P.; Rawat, T.; Bohra, M.; Trivedi, H. Effects of various concentrations of GA_3 and NAA on cuttings of hydrangea under shade net conditions. *J. Hill Agric.* **2018**, *9*, 260–264. [CrossRef]
46. Cui, Y.; Deng, Y.; Zheng, K.; Hu, X.; Zhu, M.; Deng, X.; Xi, R. An efficient micropropagation protocol for an endangered ornamental tree species (*Magnolia sirindhorniae* Noot. & Chalermglin) and assessment of genetic uniformity through DNA markers. *Sci. Rep.* **2019**, *9*, 1–10. [CrossRef]
47. Topacoglu, O.; Sevik, H.; Guney, K.; Unal, C.; Akkuzu, E.; Sivacioglu, A. Effect of rooting hormones on the rooting capability of *Ficus benjamina* L. cuttings. *Šumarski List* **2016**, *140*, 39–44. [CrossRef]
48. Asănică, A.; Tudor, V.; Sumedrea, D.; Teodorescu, R.I.; Peticilă, A.; Iacob, A. The propagation of two red and black currant varieties by hardwood cuttings combining substrate and rooting stimulators. *Sci. Papers Series B Hortic.* **2017**, *11*, 175–181.
49. Celik, H.; Cakir, S.; Celik, D.; Altun, B. Effects of leaf size and IBA on rooting and root quality of Japanese privet, rock cotoneaster and ornamental pomegranate mini-cuttings. *Acta Hortic.* **2018**, *1263*, 253–260. [CrossRef]
50. Wulandari, R.; Hasanah, Y.; Meiriani, M. Growth Response of Two Pepper (*Piper nigrum* L.) Stem Cuttings on Application of IBA (Indole Butyric Acid) and NAA (Naphthalene Acetic Acid). *Indones. J. Agric. Res.* **2018**, *1*, 87–95. [CrossRef]
51. Ada, R.; Angela, C.; Enrico, R.; Cristina, B.; Camillo, B. The weak cytokinins N, N′-bis-(1-naphthyl) urea and N, N′-bis-(2-naphthyl) urea may enhance rooting in apple and mung bean. *PCTOC* **2005**, *83*, 179–186. [CrossRef]
52. Trofimuk, L.P.; Kirillov, P.S.; Egorov, A.A. Application of biostimulants for vegetative propagation of endangered *Abies gracilis*. *J. For. Res.* **2020**, *31*, 1195–1199. [CrossRef]
53. Tikendra, L.; Amom, T.; Nongdam, P. Effect of phytohormones on rapid *In vitro* propagation of *Dendrobium thyrsiflorum* Rchb. f.: An endangered medicinal orchid. *Pharmacogn. Mag.* **2018**, *14*, 495. [CrossRef]
54. Wang, F.; Xin, X.; Wei, H.; Qiu, X.; Liu, B. In vitro regeneration, ex vitro rooting and foliar stoma studies of *Pseudostellaria heterophylla* (Miq.) Pax. *Agronomy* **2020**, *10*, 949. [CrossRef]
55. Mirani, A.A.; Abul-Soad, A.A.; Markhand, G.S. In vitro rooting of *Dendrobium nobile* Orchid: Multiple responses to auxin combinations. *Not. Sci. Biol.* **2017**, *9*, 84–88. [CrossRef]
56. Koyama, R.; Aparecido Ribeiro Júnior, W.; Mariani Zeffa, D.; Tadeu Faria, R.; Mitsuharu Saito, H.; Simões Azeredo Gonçalves, L.; Ruffo Roberto, S. Association of indolebutyric acid with *Azospirillum brasilense* in the rooting of herbaceous blueberry cuttings. *Horticulturae* **2019**, *5*, 68. [CrossRef]
57. Badawy, E.M.; El-Attar, A.B.; El-Khateeb, A.M.A. Effect of collection dates and auxins sources on rooting and growth of *Ligustrum ovalifolium* Hassk cuttings. *Plant Arch.* **2020**, *20*, 9199–9210.
58. Wang, Z.; Xu, J.; Li, H.; Yu, C.; Yin, Y. Rooting capabilities for *Taxodium* 'Zhongshanshan' 302, 118, and 405. *J. Zhejiang Univ.* **2015**, *32*, 648–654.
59. Kishore, G.R. Effect of type of cuttings and concentration of NAA on the rooting performance of jasmine (*Jasminum humile*). *HRS* **2016**, *5*, 86–87.
60. Salmi, M.S.; Hesami, M. Time of collection, cutting ages, auxin types and concentrations influence rooting *Ficus religiosa* L. stem cuttings. *J. Appl. Environ. Biol. Sci.* **2016**, *6*, 124–132.
61. Patil, Y.B.; Saralch, H.S.; Chauhan, S.K.; Dhillon, G.P.S. Effect of Growth Hormone (IBA and NAA) on Rooting and Sprouting Behaviour of *Gmelina arborea* (Roxb.). *Indian For.* **2017**, *143*, 81–85.

62. Mishra, J.P.; Bhadrawale, D.; Rana, P.K.; Mishra, Y. Evaluation of cutting diameter and hormones for clonal propagation of *Bambusa balcooa* Roxb. *J. For. Res.* **2019**, *24*, 320–324. [CrossRef]
63. Tien, L.; Chac, L.; Oanh, L.; Ly, P.; Sau, H.; Hung, N.; Thanh, V.; Doudkin, R.; Thinh, B. Effect of auxins (IAA, IBA and NAA) on clonal propagation of *Solanum procumbens* stem cuttings. *Plant Cell Biotechnol. Mol. Biol.* **2020**, *21*, 113–120.
64. Seiar, Y.A. Effect of Growth Regulators on Rooting of Cuttings in Pomegranate (*Punica granatum* L.) Cv. 'Bhagwa'. *J. Hortic. Sci.* **2017**, *11*, 156–160.
65. Hasan, A.M.; Mohamed Ali, T.J.; Al-Taey, D.K. Effects of winter foliar fertilizing and plant growth promoters on element and carbohydrate contents on the shoot of navel orange sapling. *Int. J. Fruit Sci.* **2020**, *20*, 682–691. [CrossRef]
66. Hasan, A.M.; Al-Falahy, T.H.; Al-Taey, D.K. The effect of cutting diameter and storage method on the rooting and growth of pomegranate cuttings (Salimi and Rawa cultivars). *Int. J. Agricult. Stat. Sci.* **2020**, *16*, 1457–1463.
67. Benbya, A.; Cherkaoui, S.; Gaboun, F.; Chlyah, O.; Delporte, F.; Alaoui, M.M. Clonal propagation of *Argania spinosa* (L.) skeels: Effects of leaf retention, substrate and cutting diameter. *Adv. Hortic. Sci.* **2021**, *35*, 61–72. [CrossRef]
68. Kabanova, S.A.; Musoni, W.; Zenkova, Z.N.; Danchenko, M.A.; Scott, S.A.; Kabanov, A.N. Selection of Scots pine seedling growth stimulants in extreme conditions of the Northern Kazakhstan steppe zone. In *IOP Conference Series: Earth and Environmental Science*; IOP Publishing: Bristol, UK, 2020; Volume 611, p. 012039. [CrossRef]
69. Ghosh, A.; Dey, K.; Mani, A.; Bauri, F.K.; Mishra, D.K. Efficacy of different levels of IBA and NAA on rooting of Phalsa (*Grewia asiatica* L.) cuttings. *Int. J. Chem. Stud.* **2017**, *5*, 567–571.
70. Kumar, S.; Muraleedharan, A.; Kamalakannan, S.; Sudhagar, R.; Sanjeevkumar, K. Effect of rooting hormone on rooting and survival of Nerium (*Nerium odorum* L.) var. Pink single. *Plant Arch.* **2020**, *20*, 3017–3019.

 horticulturae

MDPI

Article

Effect of Foliar Supplied PGRs on Flower Growth and Antioxidant Activity of African Marigold (*Tagetes erecta L.*)

Sadia Sadique [1,†], Muhammad Moaaz Ali [2,3,†], Muhammad Usman [4], Mahmood Ul Hasan [3], Ahmed F. Yousef [2,5], Muhammad Adnan [6], Shaista Gull [3,7] and Silvana Nicola [8,*]

1 Department of Chemistry, University of Agriculture, Faisalabad 38000, Pakistan; sadiajee81@gmail.com
2 College of Horticulture, Fujian Agriculture and Forestry University, Fuzhou 350002, China; muhammadmoaazali@yahoo.com (M.M.A.); ahmedfathy201161@yahoo.com (A.F.Y.)
3 Institute of Horticultural Sciences, University of Agriculture, Faisalabad 38040, Pakistan; mahmoodulhassan1947@gmail.com (M.U.H.); shaistagull205@gmail.com (S.G.)
4 Institute of Soil and Environmental Science, University of Agriculture, Faisalabad 38040, Pakistan; usmansomi90@gmail.com
5 Department of Horticulture, College of Agriculture, University of Al-Azhar (Branch Assiut), Assiut 71524, Egypt
6 Department of Agronomy, College of Agriculture, University of Sargodha, Sargodha 40100, Pakistan; mughal3368@gmail.com
7 Department of Horticulture, Bahauddin Zakariya University, Multan 66000, Pakistan
8 Department of Agricultural, Forest, and Food Sciences—DISAFA, Horticultural Sciences—INHORTOSANITAS, University of Torino, 10095 Grugliasco, Italy
* Correspondence: silvana.nicola@unito.it
† Equally contributed authors.

Abstract: Marigold is one of the commercially exploited flowering crops that belongs to the family Asteraceae. The production of economical yield and better quality of marigold flowers requires proper crop management techniques. Crop regulation is an important technique to make the marigold production profitable. This can be done by adopting application of plant growth regulators (PGRs). The present study was designed to investigate the effect of PGRs on flowering and antioxidant activity of two cultivars of African marigold (*Tagetes erecta L.*) viz. "Pusa Narangi Gainda" (hereinafter referred to as Narangi) and "Pusa Basanthi Gainda" (hereafter referred to as Basanthi). Plants were sprayed with abscisic acid (ABA), N-acetyl thiazolidine (NAD), gibberellic acid (GA_3), salicylic acid (SA), indole-3-butyric acid (IBA) and oxalic acid (OA) at the concentrations of 100, 150, 250, 300 and 800 mg·L^{-1}, each. Results revealed that the plants treated with 500–600 mg·L^{-1} IBA exhibited maximum increase in floral diameter (34–51%). The use of 500–550 mg·L^{-1} IBA exhibited maximal enhancement in flower fresh weight (21–92%). The exogenously applied OA significantly ($p \leq 0.05$) improved flower dry weight, total phenolic contents, total flavonoid contents and reducing power ability of marigold plants. Overall, "Narangi" performed better than "Basanthi", in terms of flowering and antioxidant activity. Conclusively, the results suggest that foliar application of PGRs favors flowering and antioxidant activity of African marigold.

Keywords: plant growth regulators; salicylic acid; oxalic acid; DPPH; antioxidant activity; reducing power ability

Citation: Sadique, S.; Ali, M.M.; Usman, M.; Hasan, M.U.; Yousef, A.F.; Adnan, M.; Gull, S.; Nicola, S. Effect of Foliar Supplied PGRs on Flower Growth and Antioxidant Activity of African Marigold (*Tagetes erecta L.*). *Horticulturae* **2021**, *7*, 378. https://doi.org/10.3390/horticulturae7100378

Academic Editor: Piotr Salachna

Received: 16 September 2021
Accepted: 6 October 2021
Published: 8 October 2021

1. Introduction

African marigold (*Tagetes erecta L.*) belongs to family Asteraceae and is one of the major and important commercial flower crops and widely grown for loose flower production [1]. It is an ornamental plant species with known medicinal use due to the high content of carotenoids and phenolics in flower petals [2]. It is popular throughout the world because of its wide spectrum of attractive colours, shape and good keeping quality. Marigold has gained popularity on account of its easy cultivation, wide adaptability and production throughout the year [3]. Apart from beautification, its flower petals are also being used for xanthophyll production which is a major carotenoid fraction and accounts for 80–90% of lutein [4,5]. The flowers are also used for religious rituals and social functions because of their wide adoptability to varying soil and climatic conditions and long duration of flowering [6].

In many countries, research was conducted to improve flowering growth of ornamental plants by treating them with environment friendly substances, e.g., gibberellic acid, oxalic acid and salicylic acid, and success was achieved to a certain level [7]. Plant growth regulators (PGRs) play an important role in flower production, which in small amounts promotes or inhibits or quantitatively modifies growth and development. Gibberellic acid has proved to be very effective in manipulating growth and flowering in chrysanthemum (*Chrysanthemum morifolium*) [8] and petunia (*Petunia hybrida*) [9]. Gibberellic acid and NAA enhance the elongation and cell division by promoting the DNA synthesis in the cell. They reduced the juvenile phase due to increase in photosynthesis and respiration with enhanced CO_2 fixation in the plant [10]. The flowering growth of marigold was reported as increased under the influence of abscisic acid [6]. Indole-3-butyric acid played a vital role in increasing floral diameter of rose species (*Rosa* spp.) by increasing cell division [11]. However, limited research was conducted in Pakistan using these PGRs to enhance growth and development of flowering plants [12].

Commercially, plant growth regulators are used for suppressing apical dominance retarding vegetative growth, lateral buds induction and the production of a large number of flowers in various crops resulting in a higher flower yield and easy cultivation [13–15]. There are many examples of utilization of plant growth hormones to regulate the flowering in aromatic plants [16]. This research is of great interest and importance for flower merchants, growers, and scientists in Pakistan. The traditionally used chemicals have negative impacts on the environment due to their nonbiodegradable characteristics [17]. Considering that there are not extensive studies about the effects of PGRs on flowering and antioxidant activity of African marigold, a pot experiment was designed to evaluate the floral growth and antioxidant response of two cultivars ("Pusa Narangi Gainda", (hereafter referred to as Narangi, and "Pusa Basanthi Gainda", hereafter referred to as Basanthi) of African marigold to exogenously applied different doses (100, 150, 250, 300 and 800 $mg{\cdot}L^{-1}$) of PGRs, including abscisic acid (ABA), N-acetyl thiazolidine (NAD), gibberellic acid (GA_3), salicylic acid (SA), indole-3-butyric acid (IBA) and oxalic acid (OA).

2. Materials and Methods

2.1. Plant Material and Experimental Site

The experiment was conducted at the research farm of the Institute of Horticultural Sciences, University of Agriculture, Faisalabad, Pakistan (31°26′03.3″ N 73°04′28.1″ E) from 16 July 2020 to 15 November 2020. One-month old seedlings of marigold cultivars "Pusa Narangi Gainda" and "Pusa Basanthi Gainda" were purchased from Qadir Bakhsh Nursery (Pvt. Ltd.), Faisalabad-38000, Pakistan, and transplanted to plastic pots (30 × 33 cm), one seedling per pot. Before transplanting, the pots were filled with coconut coir, sand and well-pulverized soil collected from the field of the research farm with the ratio of 1:3:3, respectively. After filling growing media into pots, the moisture was applied up to field capacity. The pots media had pH, EC, available phosphorus and potassium of 6.3, 0.424 dS m^{-1}, 14.92 $mg{\cdot}L^{-1}$ and 347.57 $mg{\cdot}L^{-1}$, respectively. After transplanting, pots were placed in a greenhouse. The greenhouse climate data during the complete execution of the experiment is given in Figure 1.

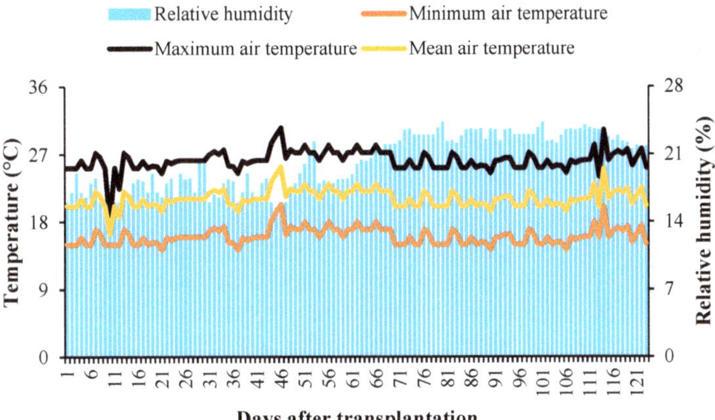

Figure 1. Microclimate conditions inside the greenhouse during the experiment (16 July to 15 November 2020) at research station of University of Agriculture, Faisalabad, Punjab, Pakistan.

2.2. PGRs Treatments

Marigold plants of both cultivars were sprayed with six different plant growth regulators (Merck KGaA, Darmstadt, Germany), namely abscisic acid (ABA), N-acetyl-thiazolidine (NAD), gibberellic acid (GA_3), salicylic acid (SA), indole-3-butyric acid (IBA) and oxalic acid (OA) at five different doses (i.e., 100, 150, 250, 300 and 800 mg·L^{-1}, each), twice a week from blooming, when each plant had ≥3 flowers. Blooming initiated in both cultivars at same time (<2 days difference). Each cultivar was sprayed with PGRs at the same time. The plants were sprayed one week after first bloom. Marigold plants were foliar sprayed with PGRs early in the morning using 1 L electronic sprayer (T Tovia, Ningbo, China) operated at a constant speed. Each treatment received 1 L of PGRs solution per spray. Control plants were treated with distilled water and maintained for comparison (0 mg·L^{-1}). Each treatment was replicated thrice, and each replication contained 10 pots, thus 10 plants.

2.3. Flowering Attributes

Flowering attributes, i.e., floral diameter, flower fresh weight and flower dry weight, were measured at full flower physiological maturity—determined by visual observation—123 days after transplanting, after 4 weeks from the first spray, that is after 8 sprays. These parameters were calculated by randomly picking 10 physiologically mature flowers from 10 plants per replicate and per treatment. Floral diameter was measured with digital Vernier callipers (DR-MV0100NG, Ningbo Dongrun Imp. & Exp. Co., Ltd., Ningbo, China), whereas flower weight was measured with digital weighing balance (MJ-W176P, Panasonic, Osaka, Japan). Fresh flowers were dried in hot air dehydrator (Ultimate 4000, Fowlers Vacola Australia Pty Ltd., Melbourne, Australia) at 65 °C for 72 h.

2.4. Antioxidant Attributes

2.4.1. Sample Preparation

The oven dried flowers were grinded and mixed in methanol to prepare sample solution (1:15 w/v). The mixture was stirred for 2 h and kept at room temperature for 24 h. Then, it was filtered and kept in sealed bottles in the dark [18].

2.4.2. Total Phenolic Contents

The 1.0 mL of each sample solution and gallic acid standard solution (20, 40, 60, 80 and 100 mg·L^{-1}), 5 mL of Folin–Ciocalteu reagent and 4 mL sodium carbonate (7% w/v) were added in a flask and shaken to mix the components completely. After keeping all the samples in the dark for 30 min, absorbance was measured at 765 nm using a spectrophotometer (T60 U Spectrophotometer, PG Instruments Ltd., Leicestershire, UK). Reagent solution was used as a blank. The amount of total phenolics was expressed as gallic acid equivalent in milligram per gram plant dry weight [19].

2.4.3. Total Flavonoid Contents

The 1.0 ml of sample or catechin standard solution (20, 40, 60, 80 and 100 mg·L^{-1}) was mixed with 4.0 mL of water in 10 mL volumetric flask followed by addition of 0.3 mL of 5% $NaNO_2$. After 5 min, 0.3 mL of 10% $AlCl_3$ was added and after waiting for one more min, 2 mL of 1 M NaOH were added, and total volume was made up to 10 mL using deionized distilled water (DDW). After mixing the solution properly, the absorbance reading was measured at 510 nm using reagent as blank. The amount of total flavonoids was expressed as catechin equivalent in milligram per gram plant dry weight [20].

2.4.4. DPPH Free Radical Scavenging Activity

The 1,1-diphenyl-2-picrylhydrazine (CAS No. 1707-75-1, ≥95% purity, Sigma-Aldrich, Milwaukee, WI, USA) scavenging activity was carried out by adding DPPH solution (1.0 mL, 0.3 M) to 2.5 mL solution of plant extract or gallic acid standard. Then samples and standards were incubated at room temperature in the dark for 20 min. Finally, absorbance was recorded at 518 nm. The control solution was prepared by adding 1.0 mL of methanol to 2.5 mL of extract solution without DPPH, while the positive control was prepared by adding 1.0 mL of DPPH solutions to 2.5 mL of gallic acid. The DPPH scavenging activity was calculated using the following expression [21].

$$\text{DPPH scavenging activity (\%)} = 100 - \left[\frac{\text{Absorbance of sample}}{\text{Absorbance of control}} \times 100\right]$$

2.4.5. Plant Reducing Power Ability

The plant extract (1.0 mL) or gallic acid standard solution (20, 40, 60, 80 and 100 mg·L^{-1}) was mixed with 2.3 mL of 0.2 M phosphate buffer (pH 6.6) and 2.5 mL of 1% potassium ferricyanide ($K_3[Fe(CN)_6]$). The mixture was incubated at 37 °C for 20 min. Then, 10% trichloroacetic acid (2.5 mL) was added to the mixture and centrifuged for 10 min at 1000 rpm, the supernatant (2.5 mL) was mixed with 2.5 mL of distilled water and 0.5 mL of 0.1% $FeCl_3$. After keeping the solution for 10 min, the absorbance was measured at 700 nm [22].

2.5. Experimental Design and Statistical Data Analysis

The experiment was designed under bi-factorial completely randomized design (Cultivar × PGRs doses) with three replicates. Collected data were submitted to analysis of variance (ANOVA) and found that cultivars and PGR doses had significant ($p \leq 0.05$) interaction, except in the case of flower diameter when treated with ABA and SA. Fisher's LSD (interaction) technique was used to compare treatment means ($p \leq 0.05$) using the analytical software package "Statistix 8.1". Principle component analysis and correlation coefficient values were determined through Pearson (n) technique using "XLSTAT ver. 2019".

3. Results

3.1. Flowering Attributes

Marigold plants of "Basanthi" exhibited more floral diameter as compared to "Narangi". In case of ABA application, marigold plants of "Basanthi" had a bigger floral diameter than "Narangi" ($p \leq 0.05$). The maximum floral diameter (5.24 cm) was observed in plants of "Basanthi" treated with 250 mg·L^{-1} ABA which was 1.34-fold greater than those of untreated plants (Figure 2A). Similarly, marigold plants of "basanthi" receiving foliar application of 100, 300 and 800 mg·L^{-1} NAD exhibited a significant ($p \leq 0.05$) increase in the floral diameter. However, for NAD application above than 250 mg·L^{-1}, a sharp decrease in the average was obtained for floral diameter, reaching 2.5 cm for the maximum dose applied (800 mg·L^{-1}) (Figure 2B). Regardless of the concentration, plants of "Basanthi" showed enhanced floral diameter under the influence of GA_3, while the plants of "Narangi" showed 28% increase in the floral diameter when treated with 800 mg·L^{-1} GA_3, as compared with control (Figure 2C). Floral diameter of both cultivars under the influence of foliar application of SA was non-significantly improved (Figure 2D). Conversely, exogenously applied IBA significantly enhanced the floral diameter of "Basanthi". Maximum floral diameter (6.25 cm) was recorded in the plants of the same cultivar receiving foliar application of 150 mg·L^{-1} IBA, followed by 800 mg·L^{-1} IBA which was 1.61 and 1.52-fold greater than those of untreated plants, respectively. The plants of "Narangi" showed non-significant but comparable results to control, except when treated with 300 and 800 mg·L^{-1} IBA (Figure 2E). In the case of OA treatment, the plants treated with 300 mg·L^{-1} OA showed maximum floral diameter (5.59 cm "Narangi", 4.95 cm "Basanthi") in both cultivars as compared to control (Figure 2F).

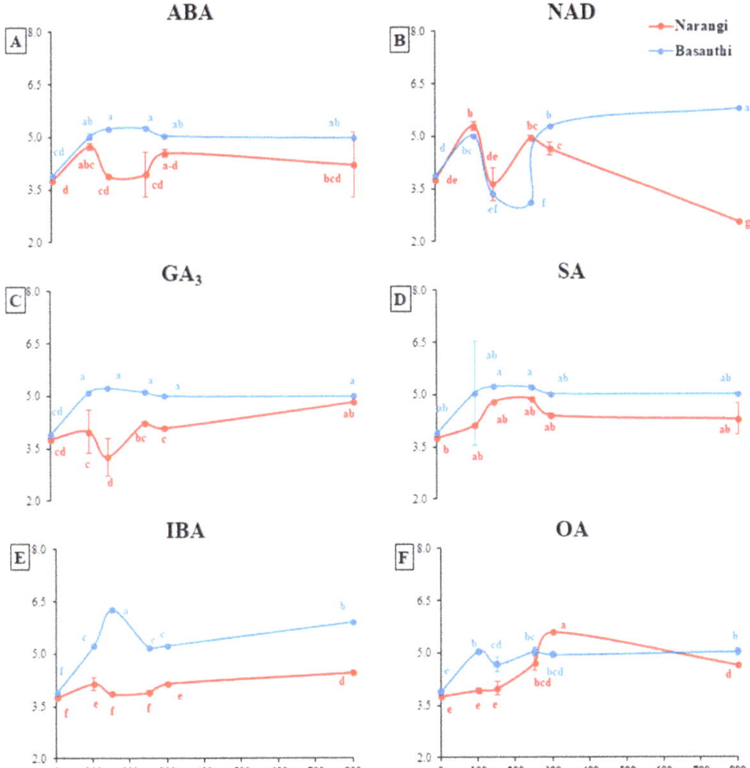

Figure 2. Floral diameter (cm) of two cultivars of marigold ("Pusa Narangi Gainda" and "Pusa Basanthi Gainda") as affected by different doses (mg·L^{-1}) of ABA (**A**); NAD (**B**); GA_3 (**C**); SA (**D**); IBA (**E**) and OA (**F**). Same letters indicate non-significant difference among treatments under Fisher's least significant difference test ($p \leq 0.05$). Vertical bars indicate average ± standard error (three replications, under bi-factorial CRD).

In case of flower fresh weight, "Narangi" showed significant response to ABA application. Plants of "Narangi" treated with 300 mg·L^{-1} ABA exhibited 62.53% increase in flower fresh weight as compared to control (Figure 3A). Under the influence of NAD, the plants of "Basanthi" receiving foliar application of 300 and 800 mg·L^{-1} NAD exhibited 85.74 and 43.88% increase in flower fresh weight, respectively. "Narangi" exhibited 93.55% increase in flower fresh weight when received 100–250 mg·L^{-1} NAD as compared to control (Figure 3B). In case of GA_3 application, "Narangi" showed maximum flower fresh weight in plants treated with 800 mg·L^{-1} (9.93 g) which was 145.26% greater than untreated plants. Meanwhile, the flower fresh weight of "Basanthi" increased up to 15% when treated with GA_3 except when treated with 300 mg·L^{-1} GA_3 (Figure 3C).

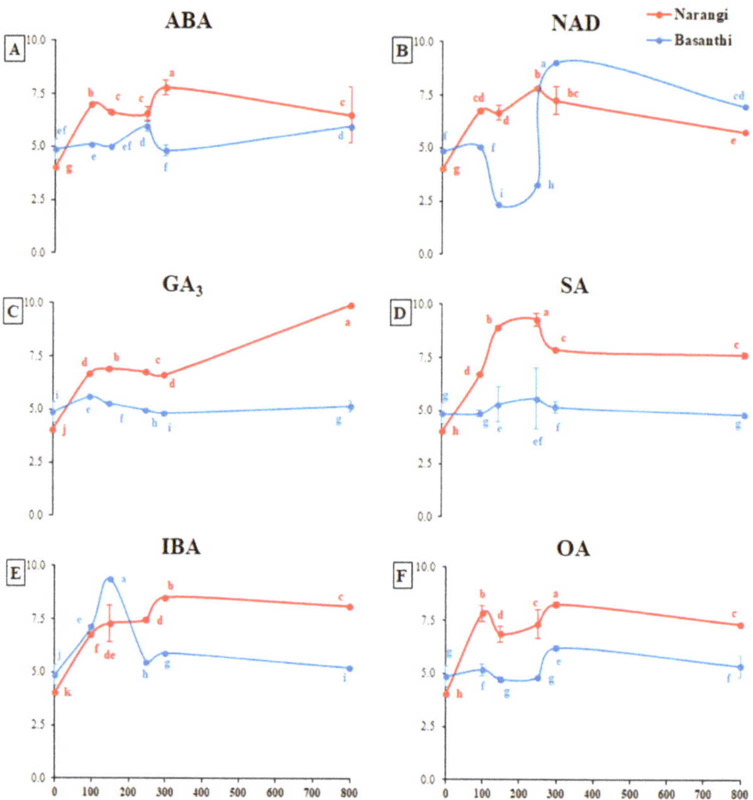

Figure 3. Flower fresh weight (g) of two cultivars of marigold ("Pusa Narangi Gainda" and "Pusa Basanthi Gainda") as affected by different doses (mg·L^{-1}) of ABA (**A**); NAD (**B**); GA_3 (**C**); SA (**D**); IBA (**E**) and OA (**F**). Same letters indicate non-significant difference among treatments under Fisher's least significant difference test ($p \leq 0.05$). Vertical bars indicate average ± standard error (three replications, under bi-factorial CRD).

The foliar application of 250 mg·L^{-1} SA showed maximum flower fresh weight (9.29 g "Narangi", 5.59 g "Basanthi") in both cultivars as compared to control (Figure 3D). Similarly, the plants of "Narangi" and "Basanthi" showed 110.23 and 92.80% increase in flower fresh weight when treated with 300 and 150 mg·L^{-1} IBA, respectively (Figure 3E). The plants treated with foliar application of 300 mg·L^{-1} OA showed maximum flower fresh weight (8.25 g "Narangi", 6.24 g "Basanthi") followed by 100 mg·L^{-1} (7.83 g "Narangi", 5.19 g "Basanthi") in both cultivars (Figure 3F).

"Basanthi" exhibited a 6.16% increase in flower dry weight when treated with 250 mg·L^{-1} ABA, while "Narangi" showed a 1.45-fold better response under the influence of 800 mg·L^{-1} ABA as compared to untreated plants (Figure 4A). The plants of "Basanthi" receiving foliar application of 100, 300 and 800 mg·L^{-1} NAD exhibited 3.90, 9.43 and 7.62% increase in flower dry weight, while the plants treated with 150 and 250 mg·L^{-1} NAD showed a

1.47 and 0.845 decrease in flower dry weight as compared to control. Flower dry weight of "Narangi" increased up to 10% with 100–300 mg·L^{-1} NAD as compared to control (Figure 4B). In case of GA_3 application, "Narangi" and "Basanthi" showed maximum flower dry weight when treated with 800 and 100 mg·L^{-1}, which were 8.15 and 3.91% more than those of untreated plants (Figure 4C). The foliar application of 150 and 250 mg·L^{-1} SA showed 12.84 and 6.61% increment in flower dry weight of "Narangi" and "Basanthi", respectively (Figure 4D). Similarly, 800 and 300 mg·L^{-1} IBA induced 6.09 and 6.47% increase in flower dry weight of the plants of "Narangi" and "Basanthi", respectively (Figure 4E). The plants treated with 300 mg·L^{-1} OA exhibited maximum flower dry weight (1.41 g "Narangi", 0.84 g "Basanthi") (Figure 5F).

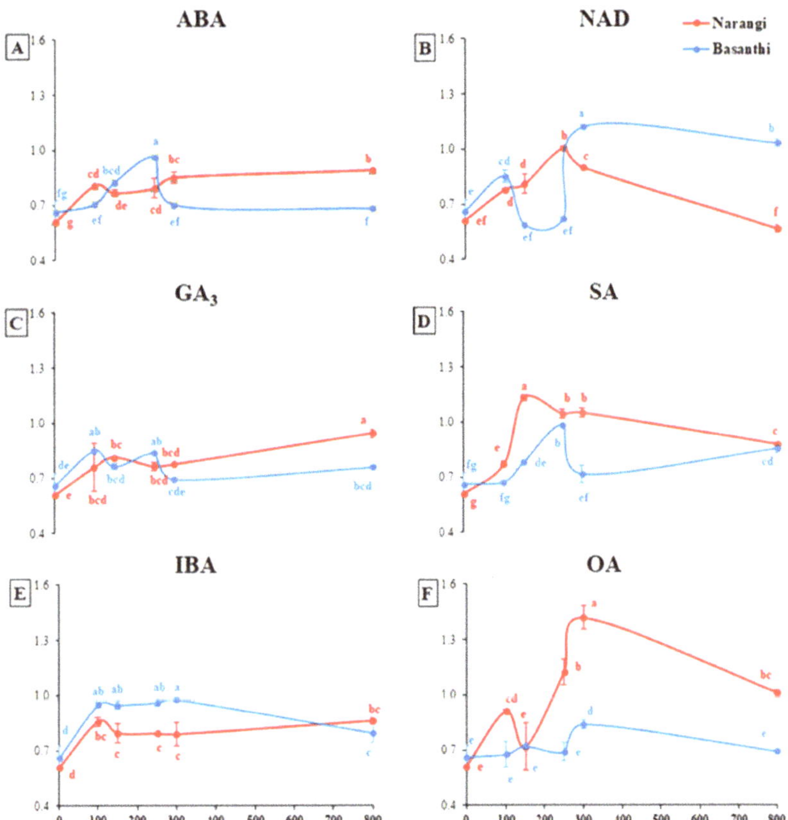

Figure 4. Flower dry weight (g) of two cultivars of marigold ("Pusa Narangi Gainda" and "Pusa Basanthi Gainda") as affected by different doses (mg·L^{-1}) of ABA (**A**); NAD (**B**); GA_3 (**C**); SA (**D**); IBA (**E**) and OA (**F**). Same letters indicate non-significant difference among treatments under Fisher's least significant difference test ($p \leq 0.05$). Vertical bars indicate average ± standard error (three replications, under bi-factorial CRD).

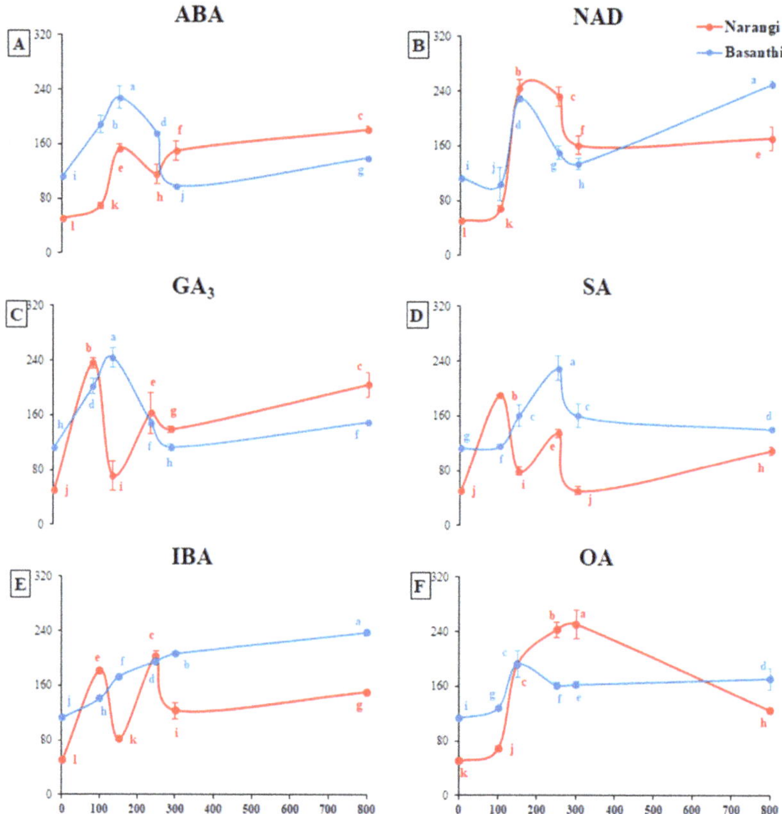

Figure 5. Total phenolic contents (mg·g^{-1}) of two cultivars of marigold ("Pusa Narangi Gainda" and "Pusa Basanthi Gainda") as affected by different doses (mg·L^{-1}) of ABA (**A**); NAD (**B**); GA_3 (**C**); SA (**D**); IBA (**E**) and OA (**F**). Same letters indicate non-significant difference among treatments under Fisher's least significant difference test ($p \leq 0.05$). Vertical bars indicate average ± standard error (three replications, under bi-factorial CRD).

3.2. Antioxidant Attributes

Marigold plants of "Basanthi" treated with 150 mg·L^{-1} ABA showed enhanced phenolic contents by 2.02-fold comparing with untreated plants (Figure 5A). Similarly, plants receiving foliar application of 150–800 mg·L^{-1} NAD exhibited up to a 121.87% increase in total phenolics. Conversely, the plants treated with 100 NAD showed decreased phenolic contents by 8.11% as compared to control (Figure 5B). "Basanthi" and "Narangi" showed an 116.36 and 368.82% increase in total phenolics when treated with 150 and 100 mg·L^{-1} GA_3, respectively (Figure 5C). Plants of "Basanthi" receiving a foliar application of 250 mg·L^{-1} SA exhibited maximum phenolic contents (229.63 mg·g^{-1}) which were 103.53% more than those of untreated plants (Figure 5D). IBA enhanced the phenolics of both cultivars in a concentration dependent manner. Maximum phenolics (237.47 mg·g^{-1}) were recorded in the plants of "Basanthi" receiving foliar application of 800 mg·L^{-1} IBA followed by the plants of same cultivar treated with 300 mg·L^{-1} IBA (Figure 5E). "Basanthi" and "Narangi" showed 397.63 and 70.46% increase in phenolics when treated with 150 and 300 mg·L^{-1} SA, respectively (Figure 5F).

The flavonoid contents of "Narangi" were recorded 69.56% more as compared to control when treated with 250 mg·L^{-1} ABA (Figure 6A). The plants of both cultivars receiving foliar application of 100, 150, 800 mg·L^{-1} NAD exhibited a significant increase in total flavonoids, while the plants of "Narangi" treated with 250 and 300 NAD showed decreased flavonoid contents as compared to the control (Figure 6B). The plants of "Basanthi" and "Narangi" treated with 300 and 100 mg·L^{-1} GA_3 showed a 2.85- and 2.04-fold increase in flavonoid contents as compared to untreated plants (Figure 6C). In case of SA application, total flavonoids of "Basanthi" significantly increased with the application of 250 and 300 mg·L^{-1} SA, while "Narangi" exhibited enhanced flavonoids as the result of 100 and 800 mg·L^{-1} SA (Figure 6D). IBA at 250 and 800 mg·L^{-1} enhanced the flavonoid contents of both cultivars by 1.89 and 2.88-fold, respectively. Maximum flavonoids (98.34 mg·g^{-1}) were recorded in the plants of "Basanthi" receiving foliar application of 800 mg·L^{-1} IBA (Figure 6E). The plants of "Basanthi" treated with 250 mg·L^{-1} OA and "Narangi" treated with 150 mg·L^{-1} OA showed maximum level of flavonoid contents (75.32 and 100.51 mg·g^{-1}, respectively) (Figure 6F).

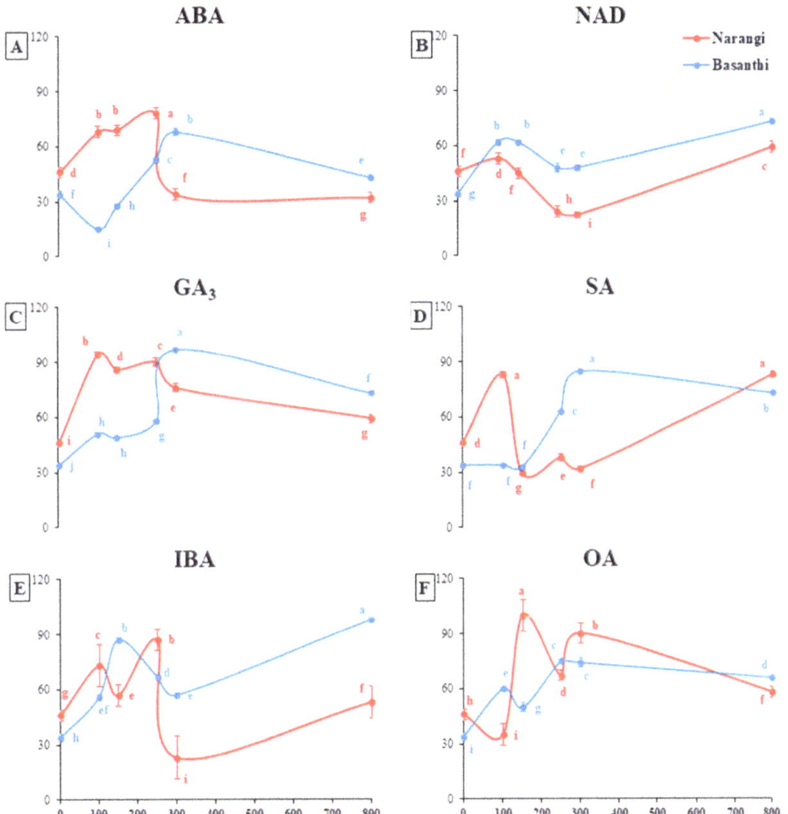

Figure 6. Total flavonoid contents (mg·g^{-1}) of two cultivars of marigold ("Pusa Narangi Gainda" and "Pusa Basanthi Gainda") as affected by different doses (mg·L^{-1}) of ABA (**A**); NAD (**B**); GA_3 (**C**); SA (**D**); IBA (**E**) and OA (**F**). Same letters indicate non-significant difference among treatments under Fisher's least significant difference test ($p \leq 0.05$). Vertical bars indicate average ± standard error (three replications, under bi-factorial CRD).

The maximum floral DPPH free radical scavenging activity (29.32%) was observed in plants of "Basanthi" treated with 800 mg·L^{-1} ABA which was 24.65% more than those of untreated plants (Figure 7A). The plants of "Narangi" treated with 100, 150 and 800 NAD, and "Basanthi" treated with 100 and 250–800 showed increase in DPPH activity up to 55 and 88%, respectively, as compared to control (Figure 7B). In case of GA_3, plants of "Narangi" showed reduction in DPPH activity, while the plants of "Basanthi" showed increased DPPH activity in a dose dependent manner (Figure 7C). The graph of SA (Figure 7D) showed opposite trend among both cultivars. The plants of "Basanthi" receiving foliar application of 800 mg·L^{-1} IBA exhibited maximum DPPH scavenging activity (35.73%) followed by the plants of "Narangi" receiving 150 mg·L^{-1} IBA (30.77%), which were 169.25 and 28.20% more than those of untreated plants of respective cultivars (Figure 7E). The exogenously applied OA also enhanced the DPPH of both cultivars except when applied at the concentration of 800 and 150 mg·L^{-1} in "Basanthi" and "Narangi", respectively (Figure 7F).

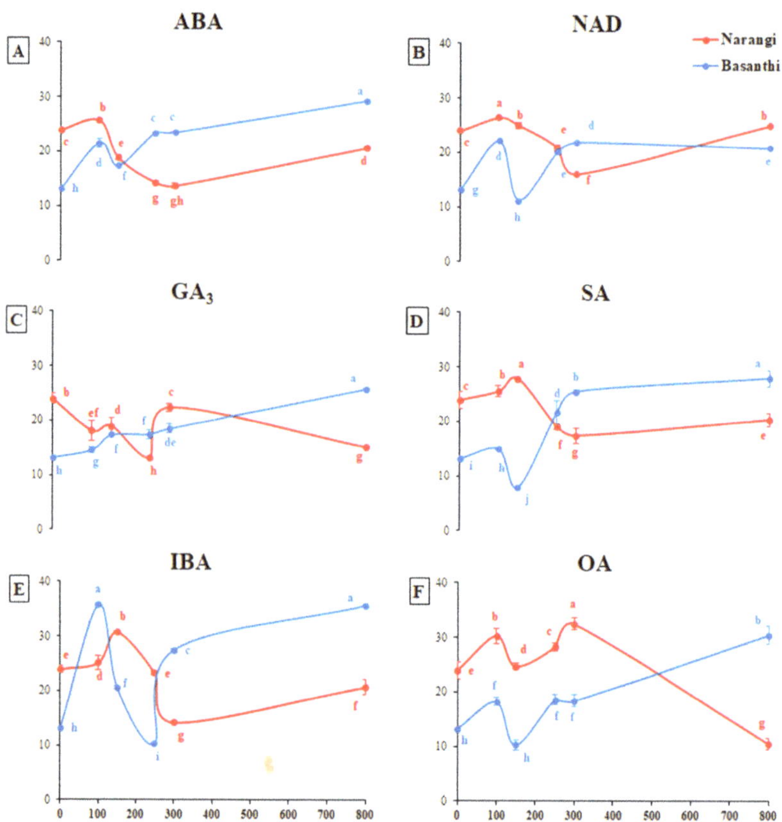

Figure 7. The DPPH free radical scavenging activity (%) of two cultivars of marigold ("Pusa Narangi Gainda" and "Pusa Basanthi Gainda") as affected by different doses of (mg·L^{-1}) ABA (**A**); NAD (**B**); GA_3 (**C**); SA (**D**); IBA (**E**) and OA (**F**). Same letters indicate non-significant difference among treatments under Fisher's least significant difference test ($p \leq 0.05$). Vertical bars indicate average ± standard error (three replications, under bi-factorial CRD).

ABA enhanced the reducing power ability (RPA) of the plants of "Basanthi" by 1.67-fold when treated with 100 mg·L^{-1} (Figure 8A). Similarly, plants receiving foliar application of NAD exhibited 1.6-times more RPA as compared to control (Figure 8B). In the case of GA_3 treatment, plants of "Basanthi" treated with 300 mg·L^{-1} and plants "Narangi" treated with 100 mg·L^{-1} showed maximum RPA (84.72 and 84.45 mg·g^{-1}, respectively), which were 59.03 and 44.48% more than those of untreated plants (Figure 8C). The RPA of both cultivars under the influence of foliar application of SA was significantly improved. Plants of "Basanthi" receiving foliar application of 250 mg·L^{-1} SA exhibited a 59.21% increase in RPA as compared to the control, which was maximum among all other dose of SA (Figure 8D). The exogenously applied IBA enhanced the RPA of both cultivars except when treated at the concentration of 800 mg·L^{-1}. Maximum RPA (84.09 mg·g^{-1} "Narangi", 78.83 mg·g^{-1} "Basanthi") was recorded in the plants receiving a foliar application of 150 mg·L^{-1} IBA (Figure 8E). In the case of OA treatment, "Basanthi" and "Narangi" were treated with 300 and 150 mg·L^{-1}, respectively, showed maximum RPA (77.23 and 80.36 mg·g^{-1}, respectively) (Figure 8F).

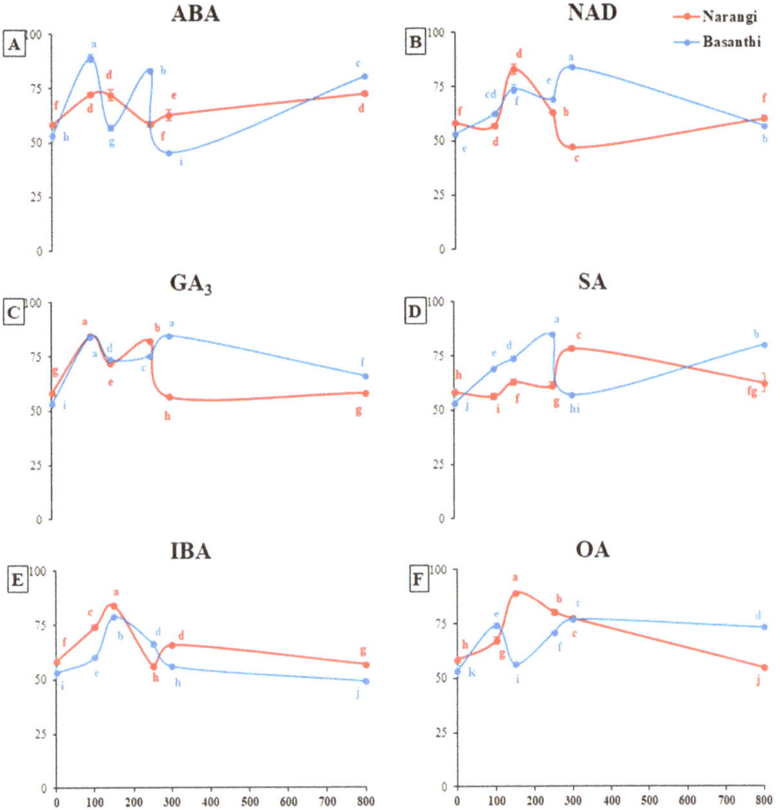

Figure 8. The reducing power ability (mg·g^{-1}) of two cultivars of marigold ("Pusa Narangi Gainda" and "Pusa Basanthi Gainda") as affected by different doses of (mg·L^{-1}) ABA (**A**); NAD (**B**); GA_3 (**C**); SA (**D**); IBA (**E**) and OA (**F**). Same letters indicate non-significant difference among treatments under Fisher's least significant difference test ($p \leq 0.05$). Vertical bars indicate average ± standard error (three replications, under bi-factorial CRD).

3.3. Principle Component Analysis

The efficacy of PGRs to modulate plant physiology depends on its concentration, application method and plant genetics [15]. Results from this study also showed that the response of marigold flower growth and antioxidant capacity to PGRs application changed with a change in the concentration and type of hormone. So, principal component analysis was conducted to delineate the concentration and PGR-dependent effects (Figure 9). Based on the highest squared cosine value corresponding to factors F1, F2 or F3, flower growth and quality attributes were clustered around PGR treatments. Factor F1, covering

20.23% variability in data (eigenvalue 2.833), showed clustering of flower diameter, flower fresh weight, flower dry weight and DPPH activity of "Basanthi" with 150–300 mg·L^{-1} NAD, 100 mg·L^{-1} SA, 150 mg·L^{-1} IBA, 300 mg·L^{-1} IBA and 150 mg·L^{-1} OA suggesting a positive influence of these treatments on these parameters. The second factor, covering 18.81% variability in data (eigen value 2.634), showed clustering of flower diameter, flower fresh weight and flower dry weight of "Narangi" with 150 mg·L^{-1} ABA, 800 mg·L^{-1} NAD, 150 mg·L^{-1} GA$_3$, 800 mg·L^{-1} GA$_3$ and 800 mg·L^{-1} OA. However, the distribution of clusters in two distinct groups on opposite sides of F1 axis indicated that application of the aforementioned PGRs had strong negative correlations with flower diameter, flower fresh weight and flower dry weight of "Narangi". The third factor of principal component analysis, covering 13.41% variability in data (eigenvalue 1.877; not shown), showed clustering of total phenolics and reducing power ability of "Narangi" with foliar application of 300 mg·L^{-1} ABA, 100 mg·L^{-1} NAD, 100 mg·L^{-1} GA$_3$ and 250–300 mg·L^{-1} OA. Thus, principal component analysis helped to delineate individual roles of PGRs with respect to their concentrations in regulating various aspects of flower growth and antioxidant attributes of African marigold.

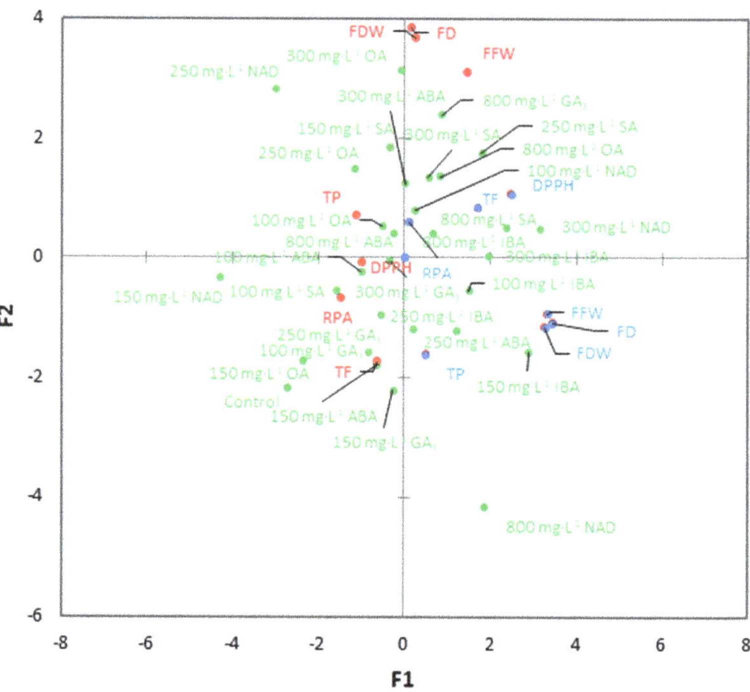

Figure 9. Principal component analysis among PGRs treatments and various flower growth and antioxidant attributes of two cultivars of marigold ("Pusa Narangi Gainda" and "Pusa Basanthi Gainda"). The treatments are shown in green colour, while variables of "Narangi" and "Basanthi" are indicated by red and blue coloured labels, respectively. Abbreviations: FD—flower diameter; FFW—flower fresh weight; FDW—flower dry weight; TP—total phenolics; TF—total flavonoids; DPPH—DPPH free radical scavenging activity; RPA—reducing power ability.

4. Discussion

4.1. Flowering Attributes

Plant growth regulators play an important role in flower production, which in small amounts promotes or inhibits or quantitatively modifies growth and development [23]. In the current study, the impact of different PGRs on floral size, weight, and some antioxidant attributes of two cultivars of African marigold, i.e., "Basanthi" and "Narangi" was evaluated. In terms of floral diameter, both cultivars differentially responded to different PGRs. "Basanthi" exhibited a 60% increase in floral diameter under the influence of 150 $mg \cdot L^{-1}$ IBA, while "Narangi" showed its maximum response (48% enhancement) under 300 $mg \cdot L^{-1}$ OA. ABA, NAD, GA_3, and SA also increased the floral diameter of marigold up to 35, 48, 34 and 34%, respectively (Figure 2). Riaz et al. [6], Mitchell and Stewart [24], and Dhuma et al. [25] reported similar results in marigold and tuberose under the influence of ABA and NAD. The GA_3-induced increment in floral diameter might be due to more synthate translocation to only a fewer sink. A similar effect of GA_3 was reported earlier in marigold and chrysanthemum [26–30]. The increase in floral diameter under the influence of SA was stated as the result of increased CO_2 assimilation, photosynthetic rate and mineral uptake as supported by previous studies [31]. Some researchers reported similar findings in calendula and marigold [32,33]. IBA plays a vital role in increasing cell division [11], and hence found a promising way to improve floral diameter of marigold. Our results about the influence of IBA and OA on floral diameter of marigold are supported by previous findings in marigold, red firespike (*Odontonema strictum*), and henna (*Lawsonia Inermis*) [34–37].

It was reported earlier that ABA is consistently effective at reducing water loss and increasing flower fresh weight of bedding plants [38]. NAD represents one of the cornerstones of cellular oxidation and is essential for plant growth and development [39]. In our findings, "Narangi" exhibited 145% more fresh weight while receiving 800 $mg \cdot L^{-1}$ GA_3, as compared to untreated plants, 57% more than the maximum observed in the plants of "Basanthi". The plants of both cultivars treated with PGRs exhibited more flower fresh weight as compared to untreated plants (Figure 3). The increase in fresh weight of marigold flowers with GA_3 might be due to the production of bigger sized flowers. This might be attributed to rapid synthesis in the cell, increase in cell size, cell elongation and rapid translocation of assimilates to sink under the influence of phytohormones [40]. SA is an emerging plant growth regulator that acts as signaling molecule in plants under biotic and abiotic stresses in marigold. SA also exerts a stimulatory effect on different physiological processes of plant growth [41]. In case of IBA treatment, the plants of both cultivars treated with 150 $mg \cdot L^{-1}$ IBA exhibited maximum flowers fresh weight. The reason behind the increase in flowers' fresh weight might be the enhancement of photosynthesis and maximum accumulation of photosynthates due to IBA application [42]. Similar results were also reported in marigold [35,43,44].

In terms of flower dry weight, "Narangi" responded better to PGRs as compared to "Basanthi". The maximum flower dry weight under the influence of PGRs was observed in the plants of "Narangi" when treated with 300 $mg \cdot L^{-1}$ OA. The plants of "Basanthi" receiving similar dose of NAD exhibited 69% more flower dry weight as compared to those of untreated plants. Other applied PGRs i.e., ABA, GA_3, SA and IBA induced a significant increase in flower dry weight of marigold up to 45, 53, 84 and 47%, respectively (Figure 4). The results about ABA and GA_3-induced floral dry weight are supported by previous findings, stating the role of PGRs in reducing water loss in the plants [28,45–49]. The results about IBA-induced flower dry weight was supported by Pacheco et al. [50] and Choudhary et al. [51] in marigold, and Pal [52] in calendula. Moreover, it was also reported that dry weight is well known to exhibit a high positive correlation with fresh weight of marigold flowers [53,54].

4.2. Antioxidant Attributes

Phenolic compounds are the secondary metabolites acting as antioxidants due to their hydroxyl group [55]. PGRs play an important role in improving antioxidant capacity of the plants. ABA was reported as having a key role in the enhancement of antioxidant capacity, anthocyanins and phenolic content of bedding plants (i.e., *Impatiens walleriana*, *Pelargonium hortorum*, *Petunia hybrida*, *Tagetes patula*, *Salvia splendens*, and *Viola wittrockiana*) [38]. In the current study, the total phenolics (floral extract) of both cultivars of *Tagetes erecta* were significantly influenced by different PGRs. Moreover, "Narangi" responded better to PGRs as compared to "Basanthi". The maximum phenolics were observed in the plants of "Narangi" and "Basanthi" under the influence of 300 mg·L^{-1} OA and 800 mg·L^{-1} NAD (Figure 5).

The effect of different PGRs on total flavonoid contents of floral extract of *Tagetes erecta* was found significant. ABA, NAD, GA$_3$, SA, IBA and OA increased floral flavonoids of marigold up to 70, 114, 185, 114, 188 and 120%, respectively. Marigold plants treated with maximum dose (800 mg·L^{-1}) of PGRs found having maximum flavonoids among other PGR doses (Figure 6). The increase in flavonoid contents was earlier reported in *Taraxacum officinale* [56] and *Zinnia elegans* [57] under the influence of GA$_3$ and SA application, respectively. IBA stimulated the production of total flavonoids in *Thymus vulgaris* and *Origanum vulgare*, but decreased it in *Ocimum basilicum* [58,59]. Likewise, OA-induced flavonoid contents were reported in *Bellis perennis* [60].

The DPPH free radical scavenging activity is considered as an acceptable mechanism to evaluate the antioxidant activity of plants [61]. The plants of "Basanthi" showed a variation of DPPH activity from the lowest value (7%) to the highest value (35.8%). The plants treated with 100 and 800 mg·L^{-1} IBA showed the highest value of free radical scavenging activity. Similarly, the plants of "Narangi" also showed varied DPPH activity from the lowest value (10%) to the highest value (35.8%). The plants of "Narangi" treated with 300 mg·L^{-1} OA exhibited maximum DPPH activity (Figure 7). Previously, it was reported that methanolic extract of *Tagetes erecta* exhibited 15.63 to 95.34% DPPH scavenging activity [62]. The antioxidant activity of plant extracts was found strongly associated with the reducing power of bioactive compounds [63]. Thus, the reducing power ability of marigold plants were determined. Our results suggest that the exogenously supplied PGRs significantly improved the reducing power ability of both cultivars (Figure 8). Similar findings in marigold were reported by some researchers [62,64].

5. Conclusions

PGRs play an important role in flower production, which in small amount promotes or inhibits or quantitatively modifies growth and development. In the present study, foliar application with different concentrations (0, 100, 150, 250, 300 and 800 mg·L^{-1}) of ABA, NAD, GA$_3$, SA, IBA and OA proved to be successful for enhancing flowering and antioxidant activity of two cultivars of African marigold. This was evidenced by the improved diameter, fresh weight, dry weight, phenolics, flavonoids, DPPH free radical scavenging activity and reducing power ability of flowers, in which the doses were increased to a certain extent, after then a detrimental effect, although not lethal, was registered. Since foliar applications of PGRs differentially regulate distinct aspects of flowers growth, specific concentrations of PGRs may help to achieve some specific quality objectives of the flowers, commercially valuable. Among the foliar applied PGRs, 150 mg·L^{-1} IBA proved to be superior in terms of maximum floral diameter of "Basanthi", whereas the maximal flower fresh weight was recorded in the plants of "Narangi" receiving foliar application of 800 mg·L^{-1} GA$_3$. Regardless of the concentration, OA significantly improved flower dry weight, total phenolic contents, total flavonoid contents and reducing power ability of "Narangi". The plants of "Basanthi" treated with 100 mg·L^{-1} IBA exhibited maximum DPPH free radical scavenging activity. The highest dose of the PGRs viz. 800 mg·L^{-1} was applied to evaluate its lethal effects on plant health. The results proved that this dose was not harmful for plants, indicating the African marigold as a hardy plant. Overall, PGRs at specific concentrations may be used as an effective exogenous application strategy to improve the flowering and antioxidant capacity of marigold flowers.

Author Contributions: Conceptualization, S.S., M.M.A. and S.N.; methodology, S.S. and M.U.; statistical analysis, M.M.A. and A.F.Y.; data curation, M.M.A.; writing—original draft preparation, S.S. and M.M.A.; writing—review and editing, M.U.H., M.A., S.G. and S.N. All authors have read and agreed to the published version of the manuscript.

Funding: This research received no external funding.

Institutional Review Board Statement: Not applicable.

Informed Consent Statement: Not applicable.

Acknowledgments: Authors would like to thank all field technicians at Institute of Horticultural Sciences, University of Agriculture, Faisalabad.

Conflicts of Interest: The authors declare no conflict of interest.

References

1. Bussmann, R.W.; Batsatsashvili, K.; Kikvidze, Z.; Paniagua-Zambrana, N.Y.; Khutsishvili, M.; Maisaia, I.; Sikharulidze, S.; Tchelidze, D. *Tagetes erecta L.* Asteraceae. In *Ethnobotany of the Mountain Regions of Far Eastern Europe;* Springer: Berlin/Heidelberg, Germany, 2020; pp. 1–5.
2. Parađiković, N.; Teklić, T.; Zeljković, S.; Lisjak, M.; Špoljarević, M. Biostimulants research in some horticultural plant species—A review. *Food Energy Secur.* **2019**, *8*, e00162. [CrossRef]
3. Choudhary, M.; Beniwal, B.S.; Kumari, A. Characterization of marigold genotypes using morphological characters. *Res. Crop.* **2014**, *15*, 839. [CrossRef]
4. Afzal, I.; Rahim, A.; Qasim, M.; Younis, A.; Nawaz, A.; Bakhtavar, M.A. inducing salt tolerance in french marigold (Tagetes patula) through seed priming. *Acta Sci. Pol. Hortorum Cultus* **2017**, *16*, 109–118. [CrossRef]
5. Singh, R.; Meena, M.L.; Verma, S.; Vilas, R.; Saurabh, V.; Mauriya, S.K.; Maurya, R.; Kumar, M. A review on performance of gibberellic acid on African marigold. *J. Pharmacogn. Phytochem.* **2019**, *8*, 43–46.
6. Riaz, A.; Younis, A.; Taj, A.R.; Karim, A.; Tariq, U.; Munir, S.; Riaz, S. Effect of drought stress on growth and flowering of marigold (*Tagetes erecta* L.). *Pak. J. Bot.* **2013**, *45*, 123–131.
7. Jowkar, M.; Salehi, H. Effects of different preservative solutions on the vase life of cut tuberose flowers at usual home conditions. *Acta Hortic.* **2005**, 411–416. [CrossRef]
8. Gautam, S.K.; Sen, N.L.; Jain, M.C.; Dashora, L.K. Effect of plant growth regulators on growth, flowering and yield of chrysanthemum (Chrysanthemum morifolium Ram.) cv. 'Nilima. *Orrisa J. Hortic.* **2006**, *34*, 36–40.
9. Sharma, L.; Collis, J.P. Effect OF plant growth regulators on growth and flower yield of petunia (Petunia hybrida L.). *J. Plant. Dev. Sci.* **2017**, *9*, 513–514.
10. Murugan, V.T.; Manivannan, K.; Nanthakumar, S. Studies on the effect of plant growth regulators on growth, flowering and xanthophyll content of African marigold (*Tagetes erecta* L.). *Int. J. Curr. Microbiol. Appl. Sci.* **2020**, *9*, 3767–3771. [CrossRef]
11. Shahzad Akhtar, M.; Aslam Khan, M.; Riaz, A.; Younis, A. Response of different rose species to different root promoting hormones. *Pak. J. Agric. Res.* **2002**, *39*, 297–299.
12. Aziz, S.; Younis, A.; Jaskani, M.; Ahmad, R. Effect of PGRs on antioxidant activity and phytochemical in delay senescence of lily cut flowers. *Agronomy* **2020**, *10*, 1704. [CrossRef]
13. Pawar, A.; Saraladevi, D.; Kannan, M.; SabirAhamed, A. Effect of micronutrients and plant growth regulators on growth, flowering and yield attributes of marigold (*Tagetes erecta* L.). *Madras Agric. J.* **2019**, *106*, 4–6. [CrossRef]
14. Ali, M.M.; Yousef, A.F.; Li, B.; Chen, F. Effect of environmental factors on growth and development of fruits. *Trop. Plant Biol.* **2021**, 1–13. [CrossRef]
15. Ali, M.M.; Anwar, R.; Malik, A.U.; Khan, A.S.; Ahmad, S.; Hussain, Z.; Hasan, M.U.; Nasir, M.; Chen, F. Plant growth and fruit quality response of strawberry is improved after exogenous application of 24-epibrassinolide. *J. Plant Growth Regul.* **2021**, *40*, 1–14. [CrossRef]
16. Ibrahim, M.; Agarwal, M.; Yang, J.O.; Abdulhussein, M.; Du, X.; Hardy, G.; Ren, Y. Plant growth regulators improve the production of volatile organic compounds in two rose varieties. *Plants* **2019**, *8*, 35. [CrossRef]
17. Ali, S.; Charles, T.; Glick, B. Delay of flower senescence by bacterial endophytes expressing 1-aminocyclopropane-1-carboxylate deaminase. *J. Appl. Microbiol.* **2012**, *113*, 1139–1144. [CrossRef]
18. Bhattacharyya, S.; Datta, S.; Mallick, B.; Dhar, P.; Ghosh, S. Lutein content and in vitro antioxidant activity of different cultivars of indian marigold flower (Tagetes patula L.) extracts. *J. Agric. Food Chem.* **2010**, *58*, 8259–8264. [CrossRef]
19. Vatai, T.; Škerget, M.; Knez, Ž. Extraction of phenolic compounds from elder berry and different grape marc varieties using organic solvents and/or supercritical carbon dioxide. *J. Food Eng.* **2009**, *90*, 246–254. [CrossRef]

20. Li, W.; Gao, Y.; Zhao, J.; Wang, Q. Phenolic, flavonoid, and lutein ester content and antioxidant activity of 11 cultivars of chinese marigold. *J. Agric. Food Chem.* **2007**, *55*, 8478–8484. [CrossRef] [PubMed]
21. Ramadan, M.F.; Kroh, L.W.; Mörsel, J.-T. Radical scavenging activity of black cumin (Nigella sativa L.), coriander (Coriandrum sativum L.), and niger (Guizotia abyssinica Cass.) crude seed oils and oil fractions. *J. Agric. Food Chem.* **2003**, *51*, 6961–6969. [CrossRef] [PubMed]
22. Habila, J.D.; Bello, I.A.; Dzikwi, A.A.; Musa, H.; Abubakar, N. Total phenolics and antioxidant activity of Tridax procumbens Linn. *Afr. J. Pharm. Pharmacol.* **2010**, *4*, 123–126. [CrossRef]
23. Kumar, M.; Singh, A.K.; Kumar, A. Effect of plant growth regulators on flowering and yield attributes of African marigold (*Tagetes Erecta* L.). *Plant. Arch.* **2014**, *14*, 363–365.
24. Mitchell, J.W.; Stewart, W.S. Comparison of growth responses induced in plants by naphthalene acetamide and naphthalene acetic acid. *Bot. Gaz.* **1939**, *101*, 410–427. [CrossRef]
25. Dhumal, S.; Kaur, M.; Dalave, P.; Garande, V.K.; Pawar, R.D.; Ambad, S.S. Regulation of growth and flowering in tuberose with application of bio-regulators. *Int. J. Curr. Microbiol. Appl. Sci.* **2018**, *7*, 1622–1626. [CrossRef]
26. Shinde, K.H.; Parekh, N.S.; Upadhyay, N.V.; Patel, H.C. Investigation of different levels of gibberellic acid (GA3) and pinching treatments on growth, flowering and yield of chrysanthemum (Chrysanthemum morifolium Ramat.) cv. "IIHR-6" under middle Gujarat conditions. *Asian J. Hortic.* **2010**, *5*, 416–419.
27. Shivaprakash, B.; Hugar, A.; Kurubar, A.; Vasudevan, S.; Husain, S. Studies on impact of bio-fertilizers and GA 3 on growth and flower yield of marigold (*Tagetes erecta* L.) cv. *Asian J. Hortic.* **2011**, *6*, 406–411.
28. Pushkar, N.C.; Singh, A.K. Effect of pinching and growth retardants on flowering and yield of African marigold (*Tagetes erecta* L.) var. Pusa Narangi Gainda. *Int. J. Hortic.* **2012**, *10*, 109–111. [CrossRef]
29. Kumar, R.; Ram, M.; Gaur, G.S. Effect of GA3 and ethrel on growth and flowering of African marigold cv. Pusa Narangi Gainda. *Indian J. Hortic.* **2010**, *67*, 362–366.
30. Kumar, N.; Kumar, J.; Singh, J.; Kaushik, H.; Singh, R. Effect of GA3 and Azotobacter on growth and flowering in African marigold (*Tagetes erecta* L.) cv. Pusa Narangi Gainda. *Asian J. Hortic.* **2016**, *11*, 382–386. [CrossRef]
31. Karlidag, H.; Yildirim, E.; Turan, M. Salicylic acid ameliorates the adverse effect of salt stress on strawberry. *Sci. Agric.* **2009**, *66*, 180–187. [CrossRef]
32. Alirezaie, H.B.; Neamati, H. Impact of exogenous salicylic acid on growth and ornamental characteristics of calendula (Calendula officinalis L.) under salinity stress. *J. Stress Physiol. Biochem.* **2012**, *8*, 258–267.
33. Kumar, A.; Kumar, J.; Mohan, B.; Singh, J.; Rajbeer; Ram, N. Effect of plant growth regulators on growth, flowering and yield of African marigold (*Tagetes erecta* L.) cv. Pusa Narangi Gainda. *Asian J. Hortic.* **2011**, *6*, 418–422.
34. Kaushik, S.; Shukla, N. Effect of IBA and NAA and their combination on the shooting of stem cuttings of African marigold (*Tagetes erecta* L.) cv. Pusa Narangi Gainda. *Int. J. Chem. Stud.* **2020**, *8*, 1800–1802. [CrossRef]
35. Ullah, Z.; Abbas, S.J.; Naeem, N.; Lutfullah, G.; Malik, T.; Khan, M.A.U.; Khan, I. Effect of indolebutyric acid (IBA) and naphthaleneacetic acid (NAA) plant growth regulaters on Mari gold (*Tagetes erecta* L.). *Afr. J. Agric. Res.* **2013**, *8*, 4015–4019.
36. Rezazadeh, A.; Harkess, R.L.; Bi, G. Effect of plant growth regulators on growth and flowering of potted red firespike. *HortTechnology* **2016**, *26*, 6–11. [CrossRef]
37. Morshed, M.N.; Deb, H.; Azad, S.A.; Sultana, M.Z.; Ashaduzzaman; Guha, A.K. Aqueous and solvent extraction of natural colorants from *Tagetes Erecta* L., Lawsonia Inermis, Rosa L for coloration of cellulosic substrates. *Am. J. Polym. Sci. Technol.* **2016**, *2*, 34–39. [CrossRef]
38. Waterland, N.L.; Campbell, C.A.; Finer, J.; Jones, M.L. abscisic acid application enhances drought stress tolerance in bedding plants. *HortScience* **2010**, *45*, 409–413. [CrossRef]
39. Hunt, L.; Lerner, F.; Ziegler, M. NAD—new roles in signalling and gene regulation in plants. *New Phytol.* **2004**, *163*, 31–44. [CrossRef] [PubMed]
40. Narute, T.T.; Parulekar, Y.R.; Narute, T.K. Effect of plant growth regulators on yield and yield attributing character of marigold cv. Calcutta Marigold under Konkan conditions. *Int. J. Curr. Microbiol. Appl. Sci.* **2020**, *9*, 3998–4005. [CrossRef]
41. Basit, A. Salicylic acid an emerging growth and flower inducing hormone in marigold (Tagetes sp. L.). *Pure Appl. Biol.* **2018**, *7*. [CrossRef]
42. Hashemabadi, D.; Zaredost, F.; Ziyabari, M.B.; Zarchini, M.; Kaviani, B.; Solimandarabi, M.J.; Torkashvand, A.M.; Zarchini, S. Influence of phosphate bio-fertilizer on quantity and quality features of marigold (Tagetes erecta L.). *Aust. J. Crop. Sci.* **2012**, *6*, 1101–1109.
43. Liao, W.; Huang, G.; Yu, J.; Zhang, M.; Shi, X. Nitric oxide and hydrogen peroxide are involved in indole-3-butyric acid-induced adventitious root development in marigold. *J. Hortic. Sci. Biotechnol.* **2011**, *86*, 159–165. [CrossRef]
44. Haq, S.U.; Shah, S.T.; Khan, N.; Khan, A.; Ahmad, N.; Ali, M.; Gul, G.; Rahman, S.; Afzaal, M.; Ullah, S.; et al. Growth and flower quality production of marigold (*Tagetes erecta* L.) response to phosphorous fertilization. *Pure Appl. Biol.* **2016**, *5*, 957–962. [CrossRef]
45. Asrar, A.-W.A.; Elhindi, K.M. Alleviation of drought stress of marigold (*Tagetes erecta*) plants by using arbuscular mycorrhizal fungi. *Saudi J. Biol. Sci.* **2011**, *18*, 93–98. [CrossRef] [PubMed]
46. Kumar, R.; Kumar, A.; Kumar, A. Effect of nutrients on growth, flowering and yield of African marigold (*Tagetes erecta* L.) cv. Pusa Narangi Gainda. *Int. J. Curr. Microbiol. Appl. Sci.* **2018**, *7*, 205–209. [CrossRef]

47. Rathore, I.; Mishra, A.; Moond, S.K.; Bhatnagar, P. Studies on effect of pinching and plant bioregulators on growth and flowering of marigold (*Tagetes erecta* L.) cv. Pusa Basanti Gainda. *Progress. Hortic.* **2011**, *43*, 52–55.
48. Khobragade, R.K.; Bisen, S.; Thakur, R.S. Effect of planting distance and pinching on growth, flowering and yield of China aster (*Callistephus chinensis.*) cv. Poornima. *Indian J. Agric. Sci.* **2012**, *82*, 334–339.
49. Sailaja, S.M.; Panchbhai, D.M.; Suneetha, K. Response of China aster varieties to pinching for growth, yield and quality. *HortFlora Res. Spectr.* **2013**, *2*, 366–368.
50. Pacheco, A.C.; Cabral, S.; Sabrina, É. Salicylic acid-induced changes to growth, flowering and flavonoids production in marigold plants. *J. Med. Plants Res.* **2013**, *7*, 3158–3163. [CrossRef]
51. Choudhary, A.; Mishra, A.; Bola, P.K.; Moond, S.K.; Dhayal, M. Effect of foliar application of zinc and salicylic acid on growth, flowering and chemical constitute of African marigold cv. pusa narangi gainda (Targets erecta L.). *J. Appl. Nat. Sci.* **2016**, *8*, 1467–1470. [CrossRef]
52. Pal, S. Role of plant growth regulators in floriculture: An overview. *J. Pharmacogn. Phytochem.* **2019**, *8*, 789–796.
53. Khanvilkar, M.; Kokate, K.D.; Mahalle, S. Performance of African marigold (*Tagetes erecta*) in North Konkan coastal zone of Maharashtra. *J. Maharashtra Agric. Univ.* **2003**, *28*, 333–334.
54. Rao, C.; Goud, P.V.; Reddy, K.M.; Padmaja, G. Screening of African marigold (*Tagetes erecta* L.) cultivars for flower yield and carotenoid pigments. *Indian J. Hortic.* **2005**, *62*, 276–279.
55. Jung, C.; Maeder, V.; Funk, F.; Frey, B.; Sticher, H.; Frossard, E. Release of phenols from Lupinus albus L. roots exposed to Cu and their possible role in Cu detoxification. *Plant Soil* **2003**, *252*, 301–312. [CrossRef]
56. Yoon, H.K.; Muhammad, H.; Abdul, L.K.; Na Chae, I.; Sang, M.K.; Hyun, H.H.; Jung, L.I. Exogenous application of plant growth regulators increased the total flavonoid content in Taraxacum officinale Wigg. *Afr. J. Biotechnol.* **2009**, *8*, 5727–5736. [CrossRef]
57. Zeb, A.; Ullah, F.; Gul, S.L.; Khan, M.; Zainub, B.; Khan, M.N. Influence of salicylic acid on growth and flowering of zinnia. *Sci. Int.* **2017**, *29*, 1329–1335.
58. Karalija, E.; Neimarlija, D.; Cakar, J.; Paric, A. Elicitation of biomass and secondary metabolite production, antioxidative and antimicrobial potential of basil and oregano induced by BA and IBA application. *Eur. J. Med. Plants* **2016**, *14*, 1–11. [CrossRef]
59. Karalija, E.; Parić, A. The effect of BA and IBA on the secondary metabolite production by shoot culture of *Thymus vulgaris* L. *Biol. Nyssana* **2011**, *2*, 29–35.
60. Siatka, T.; Kašparová, M. Seasonal variation in total phenolic and flavonoid contents and dpph scavenging activity of Bellis perennis L. flowers. *Molecules* **2010**, *15*, 9450–9461. [CrossRef]
61. Rahman, M.; Islam, B.; Biswas, M.; Alam, A.H.M.K. In vitro antioxidant and free radical scavenging activity of different parts of Tabebuia pallida growing in Bangladesh. *BMC Res. Notes* **2015**, *8*, 621. [CrossRef] [PubMed]
62. Gong, Y.; Liu, X.; He, W.-H.; Xu, H.-G.; Yuan, F.; Gao, Y.-X. Investigation into the antioxidant activity and chemical composition of alcoholic extracts from defatted marigold (*Tagetes erecta* L.) residue. *Fitoterapia* **2012**, *83*, 481–489. [CrossRef] [PubMed]
63. Bukhari, S.B.; Bhanger, M.I.; Memon, S. Antioxidative activity of extracts from fenugreek seeds (trigonella foenum-graecum). *Pak. J. Anal. Environ. Chem.* **2008**, *9*, 78–83.
64. Ćetković, G.S.; Djilas, S.M.; Čanadanović-Brunet, J.M.; Tumbas, V.T. Antioxidant properties of marigold extracts. *Food Res. Int.* **2004**, *37*, 643–650. [CrossRef]

horticulturae

MDPI

Article

Studies of Vegetative Growth, Inflorescence Development and Eco-Dormancy Formation of Abscission Layers in *Streptocarpus formosus* (Gesneriaceae)

Cherise Christina Viljoen, Muhali Olaide Jimoh and Charles Petrus Laubscher *

Department of Horticultural Sciences, Faculty of Applied Sciences, Cape Peninsula University of Technology, P.O. Box 1906, Bellville, Symphony Way, Cape Town 7535, South Africa; C.Viljoen@sanbi.org.za (C.C.V.); Jimohm@cput.ac.za (M.O.J.)

* Correspondence: LaubscherC@cput.ac.za

Abstract: *Streptocarpus formosus* (Hilliard & B.L. Burtt) T.J. Edwards is a flowering herbaceous perennial indigenous to South Africa and is part of the rosulate group of herbaceous acaulescent plants within the Gesneriaceae family. According to the National Assessment database for the Red List of South African Plants version 2020.1., the plant is listed as rare. The ornamental use of *S. formosus* has untapped commercial potential as a flowering indoor pot plant, an outdoor bedding plant for shade and as a cut flower for the vase, all of which are limited by a five-month eco-dormancy period during the late autumn and all through the cold season in the short-day winter months. Viable commercial production will require cultivation techniques that produce flowering plants all year round. This study investigated the effectiveness of applying root zone heating to *S. formosus* plants grown in deep water culture hydroponics during the eco-dormancy period in preventing abscission layer formation and in encouraging flowering and assessed the growth activity response of the plants. The experiment was conducted over eight weeks during the winter season in the greenhouse at Kirstenbosch Botanical garden in water reservoirs, each maintained at five different experimental temperature treatments (18, 22, 26—control, 30 and 34 °C) applied to 10 sample replicates. The results showed that the lowest hydroponic root zone temperature of 18 °C had the greatest effect on the vegetative growth of *S. formosus*, with the highest average increases in fresh weight (1078 g), root length (211 cm), overall leaf length (362 cm) and the number of newly leaves formed (177 = n), all noted as statistically significant when compared with the other water temperature treatments, which yielded negative results from reduced vegetative growth. Findings from the study also revealed that while all heated solutions significantly prevented the formation of abscission layers of *S. formosus*, they had a less significant effect on inflorescence formation, with only 18 °C having the greatest positive effect on flower development.

Keywords: abscission; cape primrose; eco-dormancy; flowering pot plant; hydroponics; Gesneriaceae; root zone heating; phyllomorphy; *Streptocarpus formosus*

Citation: Viljoen, C.C.; Jimoh, M.O.; Laubscher, C.P. Studies of Vegetative Growth, Inflorescence Development and Eco-Dormancy Formation of Abscission Layers in *Streptocarpus formosus* (Gesneriaceae). *Horticulturae* **2021**, 7, 120. https://doi.org/10.3390/horticulturae7060120

Academic Editor: Piotr Salachna

Received: 25 March 2021
Accepted: 28 April 2021
Published: 21 May 2021

1. Introduction

Within the Gesneriaceae, Streptocarpus form part of an economically important ornamental plant group with other significant members such as *Saintpaulia* spp. (African Violets), *Gloxinia* spp. and *Sinningia* spp. [1], all of which are herbaceous perennials known for the beauty of their flowers [2,3]. In its wild habitat in the Eastern Cape province of South Africa, *Streptocarpus formosus* flowers only in long-day, warm, summer months of the year [1,4]. *S. formosus* grows naturally in a summer rainfall locality with very little irrigation through precipitation during the cold season [4]. This abiotic combination of reduced water and low temperatures triggers a survival tactic where the nutrients and carbohydrate reserves in the leaves are transported and remobilized to actively growing parts of the plant causing yellowing of the part, or all, of the leaves [5,6] before the plants enter a survival state of eco-dormancy [7].

This annual process in *Streptocarpus* with flowering occurring mostly under 15 h long days as compared to 8 h short days [8] and combined with the slowed short-day growth processes of eco-dormancy and the shedding of leaf mass through unsightly abscission layers severely limits the ornamental commercial use of *Streptocarpus formosus*. Therefore, cultivation methods to keep plants looking attractive in active growth and to extend the flowering season are required [8,9]. The manipulation of flowering is an important aspect of the cultivation of many horticultural crops [10]. There is a high demand for flowering pot plants during all seasons, even in winter when temperatures are below optimum for flower production [11] and annual plant senescence due to low temperature causes yield reduction that results in significant economic losses to growers [12–14]. Root temperature is one of the key environmental factors that control plant growth and physiological activities in cold seasons [15,16]. Manipulating root zone temperature to keep plant crops and ornamentals actively growing for commercial out-of-season production has been comprehensively researched purposely to meet market demands [17–20].

Growth cessation, abscission formation and dormancy development, all of which are exhibited by *S. formosus*, are considerably affected by temperature [21]. Leaf loss is a physiological strategy for the avoidance of water stress in plant species adapted to drought, reducing the transpiring surface of the foliage and thereby lessening water demand [21,22]. However, leaf senescence is mainly caused by cold and less commonly by high temperature [5,23]. In *Streptocarpus formosus* with no predetermined abscission zone, leaves are either shed entirely or a 2–3 mm wide demarcation line [24] forms on the leaves and a visible difference between the basal and distal sections of the lamina is distinguishable with a dark green base and a bright yellow upper section [25]. This partial senescence in *Streptocarpus* is a perennation mechanism that ensures the protection of the basal meristem [24,26]. When the distal leaf section is completely brown and dry, this part breaks away cleanly along the abscission layer [27,28] and the leaf can continue to lengthen with new growth from the base [3,27].

The chlorophyll content is an important criterion when evaluating the ornamental value of a pot plant [29]. The degradation of chlorophyll is the cause of leaf yellowing during senescence [28,30,31]. Various studies have shown the positive effect of heated water on the retention of chlorophyll and the increased amount present within leaves [11,16,32]. The optimum temperature of the growth medium can contribute beneficially to plant physiological processes such as chlorophyll pigment formation, the accumulation of phenolic compounds and an increase in photosynthetic capacity [11].

Heat is required to increase growth to expand plant production during the cold season and can be provided with the use of greenhouses [30], and it can be combined with hydroponic growing which has become common practice to improve winter yields and to obtain optimum production under periods of suboptimal climatic conditions [11,31]. Additional benefits of heating the nutrient solution are the provision of the energy requirements for plant development, activating metabolism [32] and a reduction in pathogenic activity [33]. The application of root zone heating in a closed hydroponic system enables the volume of water to buffer temperature and contributes to energy savings compared with the expense of heating entire greenhouse structures [11,33]. With a notable global increase in the scarcity of resources and climate change, hydroponics offers workable solutions by achieving optimal growth yield and good quality crops due to the precise control of nutrition and growing conditions [17,34,35]. Yields in hydroponics average at 20–25% higher than in conventional soil cultivation and have demonstrated significantly more growth and development in root systems, which also improves the nutrient uptake ability of the plants which, in turn, leads to better shoot and leaf growth [34].

This study was designed to investigate the possibility of achieving optimum growth during the winter season by determining how the application of root zone heat could viably facilitate the ornamental production of *S. formosus*, and to evaluate the effects of different regimes of root zone temperature on abscission layers activated by eco-dormancy and the earlier formation of *S. formosus* flowers. It was also envisaged that this study would assist in determining an optimal temperature for the active growth and inflorescence formation of *S. formosus* to produce consistently high vegetative growth and flowers for cultivating superior quality pot plants in hydroponics to benefit the ornamental and floriculture industries.

2. Materials and Methods

2.1. Greenhouse Experiment

The experiment was conducted over 8 weeks during winter in the greenhouse facility at the Kirstenbosch National Botanical Garden (KNBG), Cape Town, South Africa (33°98′ 56.12″ S, 18°43′ 60.25″ E) from mid-June 2019 to mid-August 2019. Plants were grown under natural daylight conditions which provided the short-day photoperiod, 9:59:26 hr day length (15th June) to 10:54:49 hr day length (15th August), required for the experiment as *S. formosus* is then in the eco-dormancy period of its annual vegetative growth [1]. An overhead Aluminet shade net screen provided 40% shading and minimized temperature fluctuations. Maximum day temperatures ranged between 13 °C and 18 °C and night temperatures between 3 °C and 7.8 °C, with an average relative humidity between 77 and 81%.

2.2. Plant Preparation

Fifty genetically identical *S. formosus* plantlets were propagated vegetatively (Figure 1a) from one *S. formosus* mother plant. After the rooting period of four months (Figure 1b), the plantlets were thoroughly rinsed to remove the rooting media and all foreign matter from their leaves and roots. They were then potted into lattice-net plastic pots filled with 4–10 mm lightweight expanded clay aggregate (LECA) and placed in the hydroponic system with only their roots submerged in water. LECA was the preferred soilless growth medium for this study because its lightweight properties, with added porosity, would not degrade in the water while its pH remained neutral with the additional advantage of protecting the roots with its thermal insulation properties [36].

(**a**) Leaf cuttings freshly done

(**b**) Leaf cuttings matured with plantlets

Figure 1. Leaf cuttings of *S. formosus* provided n = 50 plants cultivated from one initial mother plant obtained from Kirstenbosch National Botanical Garden, Cape Town (Photos: C. Viljoen).

2.3. Hydroponic Cultivation

A closed deep water hydroponic system with an air stone and a circulating pump was used based on the recommendations, discussions and methodologies of [11,37,38]. Deepwater hydroponics allows for methods of heating the nutrient solution to the required temperature and maintains a consistent nutrient supply and temperature over the entire root surface area of the replicates. Closed hydroponics systems allow for the reuse of nutrient solution, reducing the negative environmental impacts such as leaching of fertilizers, soil and groundwater pollution, and water wastage while saving on labor. LECA provided support for the plants needing to be suspended in the nutrient solution while providing excellent aeration qualities [11,36].

Five identical deep water hydroponic systems were constructed and placed onto wire mesh tables. Each system consisted of one 70 L capacity low-density polyethylene (LDPE) reservoir filled with 60 L of aqueous nutrient solution. Each reservoir was covered with an LDPE sheet into which holes were cut to hold the 10 lattice-net (7.5 cm) plastic pots suspended (Figure 2). The pot size and depth ensured that the root zones of the plants were submerged in the nutrient solution without wetting the plant's leaf crowns, avoiding possible crown rot. To prevent oxygen deficiency and the limitations this would place on the plant growth, root aeration is essential in a hydroponic system, especially in deep water culture where there is limited air-water exchange capacity and particularly when heating the solution as there is a direct correlation between the temperature of water and the amount of oxygen it contains [35,39]. As water temperature increases, less oxygen becomes available to the roots [40], so to increase aeration all the solutions were aerated using one electromagnetic air compressor (BOYU ACQ-003) linked to each system's single air stone (50 mm), which bubbled the air up through the nutrient solution at a rate of 50 L per minute, supplying oxygen to the roots of the plants. To assist with the even distribution of both the additional air (O^2) and the heated water [37], each system's solutions were circulated using an 800 L/h hour HT submersible pump (HJ-941).

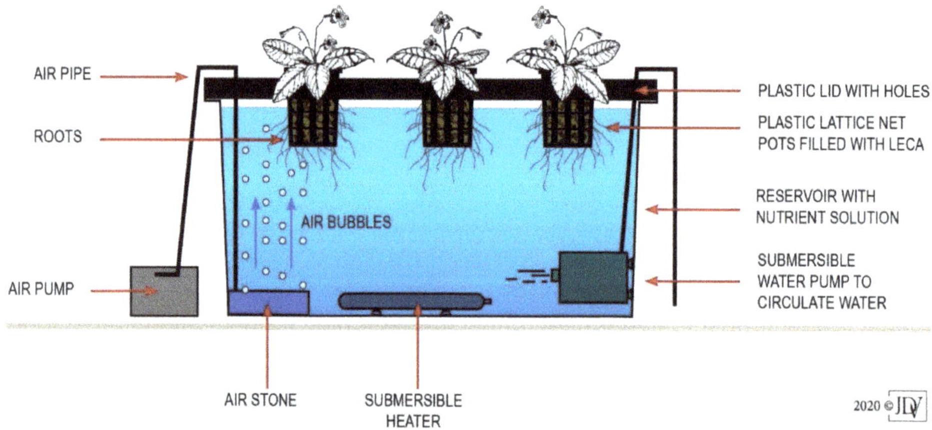

Figure 2. The closed hydroponic deep water culture system used for this study with air stone and circulating pump, and plants in lattice-net pots filled with LECA aggregate held suspended in nutrient solution (Diagram by J.D. Viljoen).

The solution comprised of ozone-treated borehole water containing Nutrifeed at a dilution rate of 1:500 (120 g in 60 L), as specified by the manufacturer Starke Ayres Pty. Ltd. Hartebeefontein Farm, Bredell Rd, Kaalfontein, Kempton Park, Gauteng, 1619, South Africa. This nutrient product supplied all the essential macro and micronutrients (6.5% Nitrogen, (N), 2.7% Phosphorous (P), 13.0% Potassium (K), 7.0% Calcium (Ca), 2.2% Magnesium (Mg), 7.5% Sulphur (S), plus Iron, Manganese, Boron, Zinc, Copper and Molybdenum) required for healthy plant growth as hydro-soluble fertilizer salts [38]. As the experiment would fall within a two-month growth period, it was decided that replacing the nutrient solution to overcome the build-up of phototoxic substances in the nutrient solution would not be required, to prevent potential disturbance damage to the roots [41,42].

The pH levels of all the nutrient solutions were monitored biweekly using a calibrated hand-held digital pH meter (HM Digital PS PH-200) and kept within a range of 6.4–7.0, a slightly acidic level recommended by [43]. The pH was adjusted accordingly using either sodium hydroxide (NaOH) to raise the pH, or hydrochloric acid (HCL) to lower the pH [42]. The various temperatures of the five test solutions were also measured for monitoring consistency. The electrical conductivity (EC) level of each system was kept within a 0.9–1.1 dSm^{-1} range as suggested for *Streptocarpus* by [44] and was used as a measure of the nutrient concentration of the solution. The EC levels and temperatures of all the nutrient solutions were monitored biweekly using a calibrated handheld (PS COM-100) EC and temperature meter produced by HM Digital Inc., Culver City, CA, USA 90230. For decreasing the EC of aqueous nutrient solutions, ozone-treated borehole water was added into reservoirs, while adding 1:500 diluted Nutrifeed™ solution increased EC levels.

2.4. Water Temperature Treatments and Experimental Design

The experiment consisted of five different hydroponic solution temperatures which were applied to 50 plants of *S. formosus* using a completely randomized block design (Figure 3; Table 1). Each temperature treatment consisted of 5 treatments with 10 replicates (n = 10), one per pot suspended in a closed deep water culture system. Pots were individually numbered and arranged randomly. The five test solutions were heated using submersible EHEIM (Plochinger Str. 54 73779 Deizisau, Germany) thermo control manually adjustable heaters as standard aquarium equipment.

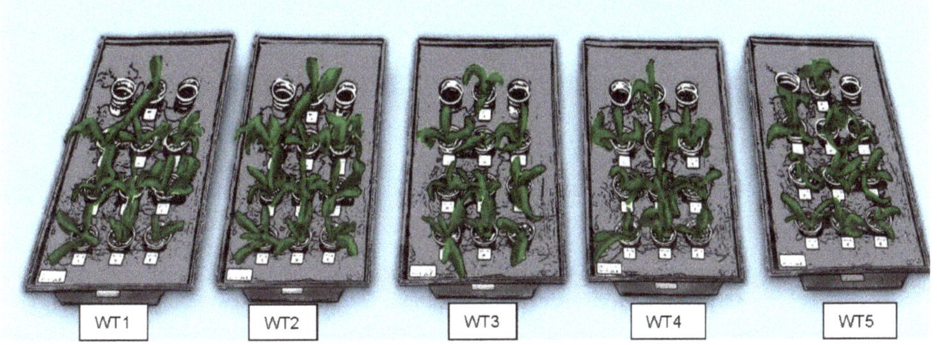

Figure 3. A completely randomized experimental block design used for the investigation (Diagram: J.D. Viljoen).

Table 1. Water temperature treatments and temperature ranges.

S/N	Treatment Code	Treatment Description
1	WT1	water temperature heated to 18 °C
2	WT2	water temperature heated to 22 °C
3	WT3	water temperature heated to 26 °C
4	WT4	water temperature heated to 30 °C
5	WT5	water temperature heated to 34 °C

The water temperature range applied was based on the ideal temperature recommendations for growing *Streptocarpus* [45], Gesneriad *Sinningia cardinalis*, [46] and other perennials [11,16], which were proven beneficial to both vegetative and inflorescence development in each case. The mean annual temperature at Port St Johns, which is *S. formosus'* natural habitat, is 19.9 °C as recorded between 1961 and 1990, with a mean summer min-max of 17.1 °C–27.6 °C when the plants are in full growth and flowering, as opposed to the mean winter min-max of 7.4 °C–20.5 °C when the plants are eco-dormant [47]. This experiment focused on applying a similar summer temperature range to the root zone to test whether the plants could thus be stimulated into active growth and flowering during the colder winter months, and WT3 at 26 °C was selected for the control as the literature reviewed indicated this to be both the ideal ambient air temperature for *Streptocarpus* under non-experimental circumstances, and a common root zone median temperature for root to shoot ratios under experimental conditions for a selection of perennial crops [11,16,32,48,49].

2.5. Vegetative Growth and Data Collection

Various measurements were taken to determine plant growth response to different nutrient solution temperatures on leaf quantity, leaf lengths, root lengths and fresh weights. Data capturing took place pre-planting and at the time of planting the plants into the quantitative research experiment system, and again post-harvest after a two-month growth period.

Before planting, each entire plantlet (roots and shoots together) was weighed for a fresh wet measurement, using an electronic laboratory scale (Sartorius Analytical Balance Scale Model type 1518) with 0.001 g readability. Additionally, at this time the lengths of both the shortest and longest leaves, as well as root length for each plantlet, were measured using a ruler [50] and recorded. The measurement of leaves was taken from the growth media level to the apex point of each leaf. All present and emerging leaves were measured, but not if less than 2 mm in length. The root length of each plant was measured by the points at which roots emerged from the stem to the tip of the root mass. Immediately after being transplanted into the LECA filled pots and placed in the deep water culture solutions, the total number of present and emerging leaves on each plantlet was then counted, but not if less than 2 mm in length, and recorded. At post-harvest, these same measurements were repeated, and the data recorded.

2.6. Inflorescence Data Collection

Immediately after being transplanted into the LECA filled pots and placed in the hydroponic test solutions, measurement of various floral parts was performed. The numbers of inflorescence stalks per plant were counted and recorded, including all present and emerging pedicels, but not if less than 2 mm in length, and the numbers of flower buds and flowers per plant were counted and recorded. After a two-month growth period, at post-harvest, these same counts were repeated and the data recorded and analyzed to determine inflorescence development in response to different nutrient solution temperatures.

2.7. Eco-Dormancy Data Collection

Immediately after being transplanted into the LECA filled pots and placed in the hydroponic test solutions, the number of abscission layers present in the leaves per plant was counted and recorded. At post-harvest, the presence or absence of abscission layers was recounted and the data recorded.

2.8. Statistical Analysis

All data collected were statistically analyzed using one-way analysis of variance (ANOVA) and computed by the software program TIBC STATISTICA Version 13.6.0. The ANOVA test was used to determine if there was a statistically significant difference between each group of water temperature's mean value. A within-between ANOVA mixed model was applied to examine for potential differences in a continuous level variable between the treatment and the control group, and over time with pre and post-tests. The occurrence of statistical difference was determined by using the Fisher Protected Least Significance Difference (L.S.D.), a pair-wise comparison technique for the comparison of two means, at values of $p < 0.05$; $p < 0.01$ and $p < 0.001$ levels of significance [51].

3. Results

3.1. Total Leaf Number

There was an interaction between hydroponic root zone temperature and the final numbers of leaves produced by the plants. The increase in the number of leaves was highly significant ($F_{1,4}$ = 34.27, $p \leq 0.0001$), and the WT1 (18 °C) treatment showed a greater increase in leaves compared to the control WT3 (26 °C) treatment by week 8 of the experiment (Table 2). The greatest increase in leaf numbers occurred at the lowest temperature treatment 18 °C, with a mean of 17.7 when compared to the 26 °C control with a mean of 6.7. Leaf numbers also displayed notably poorer results in WT2 (22 °C), mean 13.3, and WT4 (30 °C), mean 3.3. WT5 (34 °C) resulted in almost complete leaf fatality.

Table 2. The interaction of various root zone water temperatures on the overall vegetative growth of *S. formosus*.

Treatments	Temp. (°C)	ΔLeaf Number	ΔLeaf Length (cm)	ΔRoot Growth (cm)	ΔTotal Biomass (g)
WT1	18	17.7 ± 1.91 a	36.22 ± 3.75 a	21.05 ±2.28 a	107.90 ± 21.07 a
WT2	22	13.3 ± 1.30 b	15.85 ± 2.24 b	0.45 ± 0.88 c	18.67 ± 1.96 b
WT3	26	6.7 ± 0.91 d	6.68 ± 1.68 c	−5.4 ± 1.02 c	7.32 ± 1.71 b
WT4	30	3.3 ± 1.56 d	−2.18 ± 2.78 d	−23.3 ± 0.79 bc	−0.80 ± 2.16 b
WT5	34	−2.8 ± 0.99 c	−11.76 ± 2.18 e	−48.2 ± 0.41 b	−3.70 ± 1.74 b
One-way ANOVA *F*-statistic		34.2670 ***	49.0178 ***	69.6300 ***	23.7484 ***

Note: Values presented are means ± SE. The mean values followed by different letters are significantly different at $p \leq 0.001$ (***) as calculated by Fisher's least significant difference and those followed by the same letter are not significantly different.

3.2. Total Leaf Length

There was an interaction between root zone water temperature and final leaf length produced by the plants with a highly significant *F*-statistic ($F_{1,4}$ = 49.02, $p \leq 0.0001$). The 18 °C (WT1) treatment with a mean of 36.22 cm had the highest reading compared to the control 26 °C (WT3) treatment, mean 6.68 cm, or any of the other treatments by the final 8th week of the experiment (Table 2). There was thus a significant reduction in the rate of leaf length development at both WT2 22 °C (15.85 cm) and WT3 (6.68 cm), with a further reductive loss in leaf length in WT4 30 °C (−2.18 cm). This sharp decline was visually observed in the leaf health, quality and length, as temperature increases to 34 °C (WT5) compared to the control WT1 of 18 °C (Figure 4) yielded the largest increase in leaf length.

Figure 4. The visible effect of the escalating hydroponic root zone temperatures on the vegetative aerial parts of *S. formosus* evident through simple observation over the experimental period with a directly proportional reduction in leaf numbers and lengths at increasingly higher temperatures (Photos: C. Viljoen).

3.3. Total Root Growth

The statistical analysis in Table 2 indicates the greater significant values ($F_{1,4}$ = 69.63, $p \leq 0.0001$) with root growth than with the aerial parts of the plant, and indicates that WT1 (18 °C) demonstrated a notable 210.5 cm overall increase in root length, compared with only 4.5 cm at WT2 (22 °C), versus the overall negative growths of −5.4 cm for the control WT3 (26 °C) and −23.3 cm at WT4 (30 °C) treatment, with the complete death of the roots at WT5 (34 °C) treatment. Figure 5 presents the treatment interaction effect on the total root growth of the *S. formosus* plants and it indicates that heat in the root zone is a severely limiting factor when heating above a critical temperature range. Root zone temperatures at 22 °C resulted in poor root development and all the temperatures above resulted in no development and sharply declining growth or death of the *S. formosus* plant's root system.

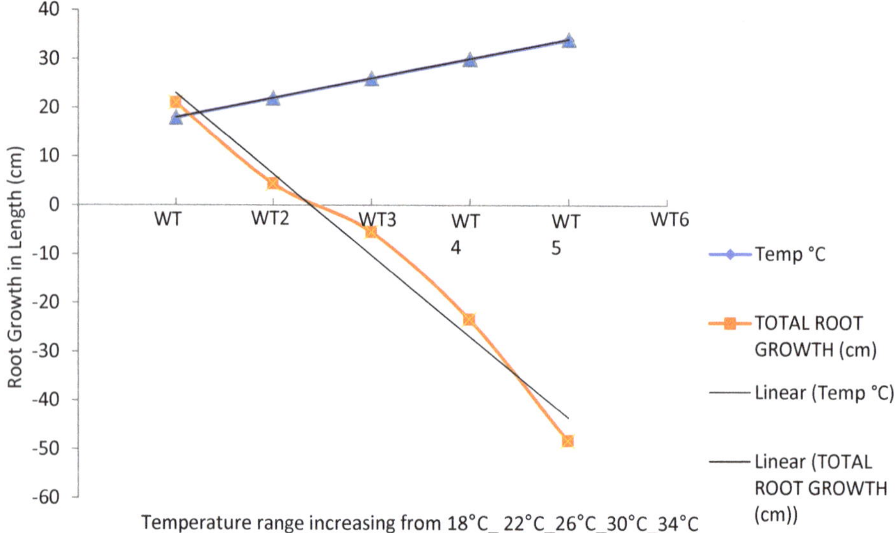

Figure 5. The relationship between root zone water temperature treatments and the relative rates of increase and decrease in root length.

3.4. Total Fresh Weight

Combined root and leaf fresh weights, as shown statistically in Table 2, were significantly affected by the root zone temperature ($F_{1,4}$ = 23.75, $p \leq 0.0001$), and the results show that incremental increases in water temperature treatments from 18 °C to 34 °C decreased fresh weight, to the point of notable fatality at the highest temperatures of 30 °C and 34 °C, clearly visible in Figure 4. The WT1 (18 °C) treatment offered the highest significant increase in overall fresh weight and vegetative growth when compared to the control WT3 (26 °C) and all the other treatments: WT2 (22 °C), WT4 (30 °C) and WT5 (34 °C). Findings from this study established that increasing hydroponic root zone solution temperature beyond the 18 °C–20 °C range did not promote the overall growth and development of *S. formosus* when compared with the significant increase in biomass growth in WT1, 18 °C, yielding a total of 107.90 g, which is equivalent to a 400% increase over 8 weeks.

3.5. Flowering in Response to Five Different Temperature Regimes in Hydroponics

The interaction between root zone heating and the inflorescence development of *S. formosus* was found to be statistically significant (Table 3), in the flower and bud formation ($F_{1,4}$ = 4.72, $p \leq 0.01$) as well as the pedicel development ($F_{1,4}$ = 4.72, $p \leq 0.001$). The highest individual mean value was evident in treatment WT1 18 °C (Figure 6); both for numbers of flowers and buds (mean 2.5) and the number of pedicels (mean 5), indicating that higher root zone temperatures WT2 (22 °C), WT3 (26 °C), WT4 (30 °C) and WT5 (34 °C) incrementally decreased inflorescence formation.

Table 3. The effect of various root zone water temperatures on the total flower development of *S. formosus*.

Treatments	Temp. (°C)	Total Number of Buds and Flowers	Total Number of Pedicels
WT1	18	2.5 ± 0.81 a	5.00 ± 1.11 a
WT2	22	1.8 ± 0.68 ab	0.90 ± 0.23 b
WT3	26	0.9 ± 0.35 cd	0.40 ± 0.22 b
WT4	30	0.1 ± 0.10 d	0.20 ± 0.13 b
WT5	34	0.0 ± 0.00 d	0.00 ± 0.00 b
One-way ANOVA *F*-statistic		4.71716 **	16.33995 ***

Values presented are means ±SE. The mean values followed by different letters are significantly different at $p \leq 0.01$ (**) and at $p \leq 0.001$ (***) as calculated by Fisher's least significant difference and those followed by the same letter are not significantly different.

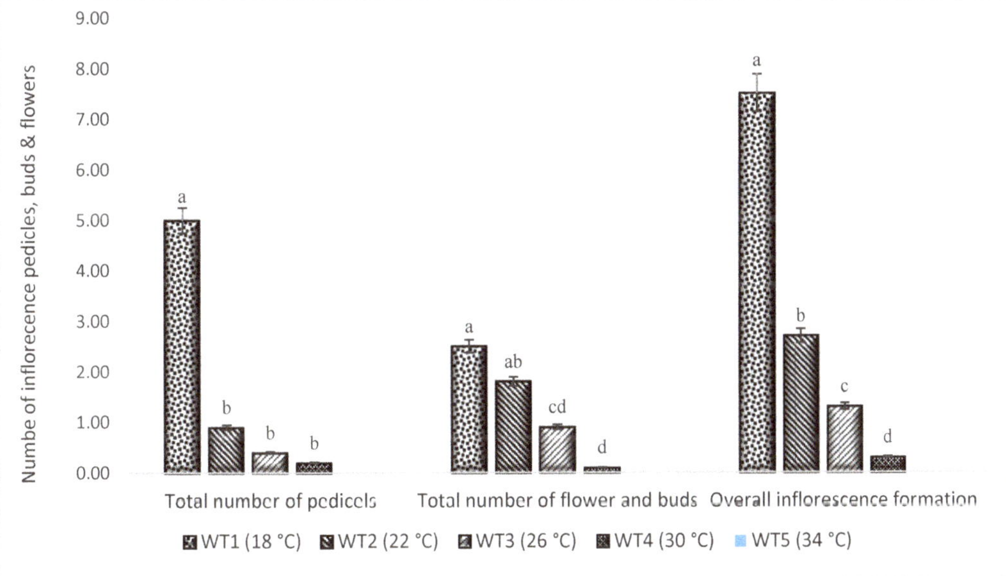

Figure 6. The relationship between root zone water temperatures and inflorescence formation. Note: The bars presented here are means ± SE. Bars with different letters are significantly different at $p \leq 0.01$ (**) (total number of flower and buds; overall inflorescence formation) and $p \leq 0.001$ (***) for the total number of pedicels as calculated by Fisher's least significant difference.

Conversely, the lowest temperature of 18 °C (WT1) significantly increased the inflorescence formation of *S. formosus*. Flowers were evident at lower root zone temperatures of 18 °C and 22 °C compared to the control treatment at 26 °C (WT3) or the higher temperatures. Increasing water temperature in the range from 26 °C to 34 °C not only decreased inflorescence formation but led to total fatality of the plants at the highest temperature of 34 °C (WT5), as seen in Figure 6. A positive finding is that at 18 °C (WT1) flowers did develop during colder short-day periods, which indicates a strong possibility that manipulating the growing temperatures could induce *S. formosus* to flower earlier in the season, thereby extending the flowering period for an all-round year commercial marketing period.

3.6. Reduction in Abscission Layers in Response to Five Different Temperature Regimes in Hydroponics

As shown in Figure 7, the effects of root zone water temperature on the reduction in abscission layers already present on the *S. formosus* replicates' leaves were statistically significant at a value ($F_{1,4}$ = 19.85, $p < 0.0005$). The few abscission layers that were present at the time of the experiment's inception all disappeared (Figure 8); however, more significantly, no abscission layers formed on any plants in the heated treatments during the winter period as would usually naturally occur (Figure 9a,b). Treatments applied in this study indicate that root zone heating is a viable method for overcoming and preventing the formation of abscission layers.

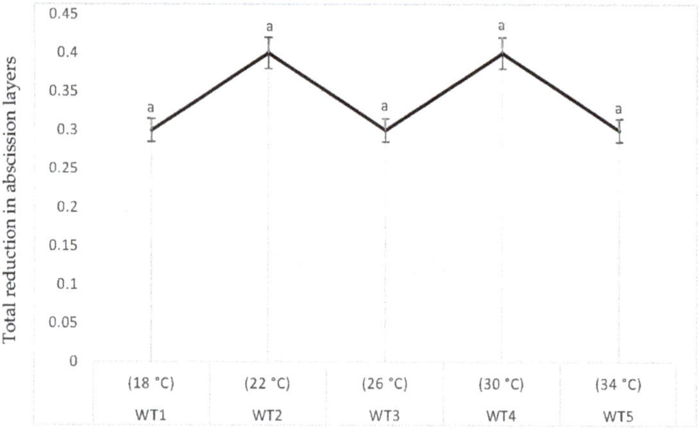

Figure 7. Root zone heating has the effect of minimizing abscission layers on *S. formosus*. Note: The line graph presented here depicts means of reduction in abscission layers ± SE. The mean values followed by the same letters are not significantly different (ns) at $p \leq 0.05$ as calculated by Fisher's least significant difference.

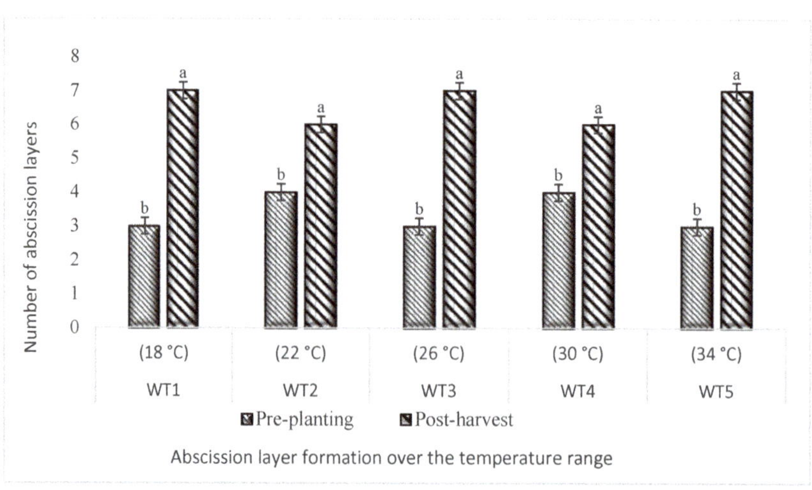

Figure 8. The correlation between all root zone water temperature treatments and the decrease in abscission layers that were present at the start, as compared to the complete absence of abscission layers at the end of the study. Note: Bar graphs presented here are means of the number of abscission layers ±SE. The mean values followed by the same letters are not significantly different (ns) at $p \leq 0.05$ as calculated by Fisher's least significant difference.

(**a**) Applied to (n) = A.

(**b**) Applied to (n) = G.

Figure 9. Postharvest photos display the effect of increasing root zone temperature across the temperature range 18 °C to 34 °C on flower and abscission layer formation on both n = A and n = G (Photos: C. Viljoen).

4. Discussion

High vegetative, flower and fruit yields in quality greenhouse crops are possible with hydroponics due to the precise control of growing conditions and required nutrients [42,52]. Nutrient solution temperature is easily controllable in hydroponics and may be manipulated to control plant growth and maximize the production of plants and flowering during winter periods [11]. Two cultivars of *Saintpaulia* (Gesneriaceae) subjected to a root zone heating range of 17 °C–25 °C exhibited a 10–15% reduced cultivation time and a significant increase in the rate of flower formation [53]. Root zone heating has shown significant results in herbaceous leafy crops, increasing flower numbers by increasing nutrient uptake [20,48]. *Chrysanthemum* responded positively when grown in a soilless culture system with a heated solution and produced flowers earlier with optimum results at 24 °C [49]. In woodier crops, such as apple, a root zone temperature of 15 °C proved to be optimal for flowering with a distinct reduction at 30 °C [54], and roses grown in a heated soilless culture system showed an increase in the number of blooms produced over the production season [55].

Streptocarpus formosus responded in various ways to different temperature regimes with a clear trend that resulted in the death of the leaves and roots at higher temperatures (26 °C to 34 °C), with the most optimal growth at the lower temperature of 18 °C. It is also clear that the roots were more sensitive than the shoots. Treatments applied in this investigation had a significant effect on the vegetative root and leaf growth as well as the overall fresh weight of *S. formosus*. The results obtained from this research disagree with various previous studies which yielded positive results in other leafy perennials and crops at higher temperature ranges such as 24 °C–28 °C for spinach [56]; 25 °C–30 °C for tomatoes and lettuce [20]; 25 °C–45 °C for muskmelon plants [57]; and 15 °C–30 °C, for *Chrysanthemum* with an optimum temperature of 24 °C [49].

Several other studies performed on soft shrubs, such as roses, indicated that shoot growth was reduced at root temperatures lower than 18 °C [55] and, at this specific temperature or above, heat in the root zone was beneficial [58]. For *Euphorbia pulcherrima* cuttings, the optimum temperature range for rooting was 25 °C–28 °C [59]. Results for conifer seedlings, such as pine (temperature range 8 °C–20 °C), had significantly new root growth at 20 °C [60]. In [19], the authors showed that lowering the temperature from 21.4 °C to 16.8 °C for *Disa* spp. had a negative effect on root growth and fresh weight, which agrees with this study where optimum vegetative growth was recorded at root temperatures lower than 18 °C.

The vegetative growth responses of *S. formosus* in this study contradict the results of research performed on cooler root zone temperature ranges, indicating that lower temperature ranges can restrict photosynthetic, respiration, metabolic and osmotic activities [20,61,62]. However, findings from this study concur with research performed on *Streptocarpus* hybrid leaf cuttings in the laboratory, which produced the most roots and buds at 12 °C and 18 °C [63], as well as with research performed on cucumbers, where the lowest temperature within the range 22 °C–33 °C yielded the best results [16].

Streptocarpus species only naturally produce flowers during the long-day summer months [3,25]. This study, however, showed that *S. formosus* was able to produce flower buds during the winter short-day period at lower temperatures. The importance of increasing flowers and regulating the timing of flowering in pot plant production can support the production of the species [61]. As confirmed in Table 3 earlier, the effect of various root zone temperatures on the total flower development of *S. formosus* was statistically significant at $p \leq 0.05$.

In *S. formosus*, the tips of the leaves often slowly die back to an abscission layer when stressed by drought, low temperature or when overwintered [3]. Growth cessation, abscission formation and dormancy development are considerably affected by temperature [21]. Leaf loss is a strategy for the avoidance of water stress in plant species adapted to drought because it reduces the transpiring surface of the foliage and therefore lessens the water demand [21,22]. Leaf senescence is mainly caused by cold and less commonly by high temperature [5,23]. In winter deciduous species, leaf senescence is an indication of the change from an active to a dormant growth stage [21,64]. When climatic conditions become unfavorable and the plants experience a state of stress, phytohormones react and leaf abscission that can lead to complete senescence is often the result [21]. Some significant abiotic factors affecting leaf abscission are nutrient availability, temperature and water supply [5,21,23], all of which can be managed within hydroponic cultivation systems [17,38]. Ref. [44] recommends that during colder months sub-irrigation should be used with minimal overhead irrigation as water that is considerably colder than the average leaf temperature causes unsightly leaf damage with yellow spots or blotches on *Streptocarpus* leaves.

In [65], the authors reported that abscission occurred due to short photoperiods in some *Streptocarpus* spp. In [66], the authors also stated that photoperiod and temperature are the main cues controlling leaf senescence in winter deciduous species, with water stress imposing an additional influence [21]. Although manipulating light and photoperiod could prevent eco-dormancy in *S. formosus* in the cold season short days, this study, however, proved that manipulating root zone temperature significantly affected the vegetative growth and flower development. A lack of water or nutrients results in leaf yellowing in many plant species, which can be reversed in some species upon removing the stress [6]. In *Streptocarpus*, it is still possible for the reversal of the formation of the abscission layer and the senescence processes even if the leaves are displaying a distal depletion of chloroplasts [65] if the plants are maintained under conditions of high temperature, nutrient levels and humidity [24]. Leaf senescence can be delayed by warming as photoperiodic triggers and growth proficiency could increase because of a slower speed or prevention of leaf senescence [21,62]. Plant growth can be controlled by the direct correlation between nutrient solution temperatures around the root zone and the uptake of nutrients, with increased plant growth at elevated root temperature correlated with higher nutrient absorption [15,67].

In this study, *S. formosus* responded in various ways to different root zone temperatures but showed the most significant vegetative mass increases in both shoots and roots at the lower temperature of 18 °C indicating that a low-temperature heating range can be used to keep *S. formosus* in active growth during the cold season. Treatments applied in this investigation had a significant effect on the flowering formation of *S. formosus*. Plants responded better to the lower root zone temperatures of 18 °C and 22 °C compared to the higher temperature intervals, and flowers were formed during the colder short-day periods, which indicates a strong possibility that manipulating the growing temperatures could induce *S. formosus* to flower earlier in the season, thereby increasing its annual commercial marketing period. These findings agree with [49] and [53], in which root zone heating increased blooms and extended the flowering period into the cold season months or encouraged earlier flowering. Moreover, treatments applied in this research also had a significant effect on the abscission layer formation of *S. formosus*. This indicates that root zone heating is a viable method for preventing the formation of abscission layers.

5. Conclusions

It is concluded that high root-zone temperatures decreased the vegetative growth of *S. formosus*. The results showed that the cooler root-zone temperature of 18 °C improved growth (leaf number, leaf and root lengths and fresh weight). This study further established that increasing root-zone temperatures did not promote the flowering of *S. formosus*; however, plants responded positively to flowering at decreased temperatures from 22 °C–18 °C. *S. formosus* has commercial potential as an indoor flowering pot plant and a flowering landscape perennial, and shows potential within the cut-flower trade. Therefore, these results will contribute to developing optimal cultivation protocols for cultivating *Streptocarpus* spp. and its hybrids and guide commercial growers in the cultivation of *S. formosus* in particular.

Author Contributions: Conceptualization, C.C.V. and C.P.L.; Data curation, C.C.V. and M.O.J.; Formal analysis, C.C.V. and M.O.J.; Funding acquisition, C.C.V. and C.P.L.; Investigation, C.C.V.; Methodology, C.C.V. and C.P.L.; Project administration, C.P.L.; Resources, C.C.V. and C.P.L.; Supervision, C.P.L.; Validation, M.O.J. and C.P.L.; Writing—original draft, C.C.V.; Writing—review & editing, C.C.V., M.O.J. and C.P.L. All authors have read and agreed to the published version of the manuscript.

Funding: The study was funded by the South African National Biodiversity Institute.

Institutional Review Board Statement: Not applicable.

Informed Consent Statement: Not applicable.

Data Availability Statement: Data is contained within the article.

Conflicts of Interest: The authors declare no conflict of interest.

References

1. Hillard, O.; Burtt, B. *Streptocarpus, an African Plant Study*; University of Natal Press: Pietermaritzburg, South Africa, 1971.
2. Otieno, M.; Joshi, N.; Rutschmann, B. Flower visitors of Streptocarpus teitensis: Implications for conservation of a critically endangered African violet species in Kenya. *PeerJ* **2021**, *9*, 1–18. [CrossRef]
3. van der Walt, L. Streptocarpus formosus. *Veld Flora* **2001**, *87*, 116–117.
4. Pooley, E. *A field Guide to Wild Flowers Kwazulu-Natal and the Eastern Region*; Natal Flora Publications Trust: Durban, South Africa, 1998.
5. Russo, S.E.; Kitajima, K. The Ecophysiology of Leaf Lifespan in Tropical Forests: Adaptive and Plastic Responses to Environmental Heterogeneity. In *Tropical Tree Physiology*; Goldstein, G., Santiago, L., Eds.; Springer: Cham, Switzerland, 2016; Volume 6, pp. 357–383.
6. Van Doorn, W.G.; Woltering, E.J. Senescence and programmed cell death: Substance or semantics? *J. Exp. Bot.* **2004**, *55*, 2147–2153. [CrossRef] [PubMed]
7. Lang, G.; Early, J.; Martin, G.; Darnell, R. Endo-, para-, and ecodormancy: Physiological terminology and classification for dormancy research. *HortScience* **1987**, *22*, 371–377.
8. White, J.W. New and renewed pot plants- Streptocarpus. *Pa. Flower Grow. Bull.* **1975**, *279*, 469–472.
9. Viljoen, C. *Streptocarpus Saxorum/Plantz Africa*; SANBI: Pretoria, South Africa, 2010.
10. Blanchard, M.G.; Runkle, E.S. Intermittent light from a rotating high-pressure sodium lamp promotes flowering of long-day plants. *HortScience* **2010**, *45*, 236–241. [CrossRef]
11. Nxawe, S.; Ndakidemi, P.A.; Laubscher, C.P. Possible effects of regulating hydroponic water temperature on plant growth, accumulation of nutrients and other metabolites. *Afr. J. Biotechnol.* **2010**, *9*, 9128–9134. [CrossRef]
12. Sade, N.; Del Mar Rubio-Wilhelmi, M.; Umnajkitikorn, K.; Blumwald, E. Stress-induced senescence and plant tolerance to abiotic stress. *J. Exp. Bot.* **2018**, *69*, 845–853. [CrossRef]
13. Mills, P.J.W.; Smith, I.E.; Marais, G. A greenhouse design for a cool subtropical climate with mild winters based on microclimatic measurements of protected environments. *Acta Hortic.* **1990**, *281*, 83–94. [CrossRef]
14. Avila-Ospina, L.; Moison, M.; Yoshimoto, K.; Masclaux-Daubresse, C. Autophagy, plant senescence, and nutrient recycling. *J. Exp. Bot.* **2014**, *65*, 3799–3811. [CrossRef]
15. Yan, Q.-Y.; Duan, Z.-Q.; Mao, J.-D.; Li, X.; Dong, F. Low Root Zone Temperature Limits Nutrient Effects on Cucumber Seedling Growth and Induces Adversity Physiological Response. *J. Integr. Agric.* **2013**, *12*, 1450–1460. [CrossRef]
16. Al-Rawahy, M.S.; Al-Rawahy, S.A.; Al-Mulla, Y.A.; Nadaf, S.K. Effect of Cooling Root-Zone Temperature on Growth, Yield and Nutrient Uptake in Cucumber Grown in Hydroponic System During Summer Season in Cooled Greenhouse. *J. Agric. Sci.* **2019**, *11*. [CrossRef]
17. Agung, P.P.; Yuliando, H. Soilless Culture System to Support Water Use Efficiency and Product Quality: A Review. *Agric. Agric. Sci. Procedia* **2015**, *3*, 283–288. [CrossRef]
18. Phantong, P.; Machikowa, T.; Saensouk, P.; Muangsan, N. Comparing growth and physiological responses of Globba schomburgkii Hook. f. and Globba marantina L. under hydroponic and soil conditions. *Emir. J. Food Agric.* **2018**, *30*. [CrossRef]
19. Pienaar, D.; Combrink, N.J.J. The effects of N-source, shading and root zone cooling on two Disa hybrids. *S. Afr. J. Plant Soil* **2007**, *24*, 166–171. [CrossRef]
20. Moorby, J.; Graves, C.J. Root and air temperature effects on growth and yield of tomatoes and lettuce. *Acta Hortic.* **1980**, 29–44. [CrossRef]
21. Estiarte, M.; Penuelas, J. Alteration of the phenology of leaf senescence and fall in winter deciduous species by climate change: Effects on nutrient proficiency. *Glob. Chang. Biol.* **2015**, *21*, 1005–1017. [CrossRef] [PubMed]
22. Kooyers, N.J. The evolution of drought escape and avoidance in natural herbaceous populations. *Plant Sci.* **2015**, *234*, 155–162. [CrossRef]
23. Addicott, F.T. Environmental Factors in the Physiology of Abscission. *Plant Physiol.* **1968**, *43*, 1471–1479.
24. Noel, A.R.A.; van Staden, J. Phyllomorph Senescence in Streptocarpus molweniensis. *Ann. Bot.* **1975**, *39*, 921–929. [CrossRef]
25. Marston, M. The morphology of a Streptocarpus hybrid and its regeneration from leaf cuttings. *Sci. Hortic.* **1964**, *17*, 114–120.
26. Rohde, A.; Bhalerao, R.P. Plant dormancy in the perennial context. *Trends Plant Sci.* **2007**, *12*, 217–223. [CrossRef]
27. Gawadi, A.G.; Avery, G.S. Leaf abscission and the so-called abscission layer. *Am. J. Bot.* **1950**, *37*, 172–180. [CrossRef]
28. Matos, F.S.; Borges, L.P.; Müller, C. Ecophysiology Of Leaf Senescence. *Agron. Agric. Sci.* **2020**, *3*, 1–6. [CrossRef]
29. Dhanasekaran, D.; Jasmine, M. Rooting Behavior of Certain Foliage Ornamentals Grown under Hydroponic Nutrient Solutions. *J. Floric.* **2019**, *21*.
30. Sethi, V.P.; Sharma, S.K. Greenhouse heating and cooling using aquifer water. *Energy* **2007**, *32*, 1414–1421. [CrossRef]
31. Grillas, S.; Lucas, M.; Bardopoulou, E.; Voulgari, M. Perlite based soilless culture systems: Current commercial application and prospects. *Acta Hortic.* **2001**, *548*, 105–113. [CrossRef]
32. Calatayud, Á.; Gorbe, E.; Roca, D.; Martínez, P.F. Effect of two nutrient solution temperatures on nitrate uptake, nitrate reductase activity, NH4+ concentration and chlorophyll a fluorescence in rose plants. *Environ. Exp. Bot.* **2008**, *64*, 65–74. [CrossRef]
33. Van Os, E.A. Recent advances in soilless culture in Europe. *Acta Hortic.* **2017**, *1176*, 1–8. [CrossRef]

34. Asao, T.; Asaduzzaman, M.; Mondal, F.M. Horticultural research in Japan. Production of vegetables and ornamentals in hydroponics, constraints and control measures. *Adv. Hortic. Sci.* **2014**, *28*, 167–178. [CrossRef]
35. Morard, P.; Lacoste, L.; Silvestre, J. Effect of oxygen deficiency on mineral nutrition of excised tomato roots. *J. Plant Nutr.* **2004**, *27*, 613–626. [CrossRef]
36. Boudaghpour, S.; Nasir, K. A study on light expanded clay aggregate (LECA) in a geotechnical view and its application on greenhouse and greenroof cultivation. *Int. J. Geol.* **2008**, *2*, 59–63.
37. Jones, J.B. *Complete Guide for Growing Plants Hydroponically*; CRS Press, Taylor and Francis Group, LLC: Abingdon, UK, 2014; ISBN 978-1-4398-7668.
38. Nkcukankcuka, M.; Jimoh, M.O.; Griesel, G.; Laubscher, C.P. Growth characteristics, chlorophyll content and nutrients uptake in Tetragonia decumbens Mill. cultivated under different fertigation regimes in hydroponics. *Crop Pasture Sci.* **2021**, *72*, 1–12. [CrossRef]
39. Sharma, N.; Acharya, S.; Kumar, K.; Singh, N.; Chaurasia, O.P. Hydroponics as an advanced technique for vegetable production: An overview. *J. Soil Water Conserv.* **2018**, *17*, 364. [CrossRef]
40. Adams, P. Crop nutrition in hydroponics. *Acta Hortic.* **1993**, 289–306. [CrossRef]
41. Kratky, B.A. A suspended pot, non-circulating hydroponic method. *Acta Hortic.* **2004**, *648*, 83–89. [CrossRef]
42. Faber, R.J.; Laubscher, C.P.; Rautenbach, F.; Jimoh, M.O. Variabilities in alkaloid concentration of *Sceletium tortuosum* (L.) N.E. Br in response to different soilless growing media and fertigation regimes in hydroponics. *Heliyon* **2020**, *6*, e05479. [CrossRef] [PubMed]
43. Martens, D. Back to Basics: Streptocarpus. *Gesnariads* **2013**, *63*, 47–50.
44. Uhl, R. Crop Culture Report: Streptocarpus Ladyslippers Series. *Greenh. Prod. News* **2012**.
45. Pennisi, B.V. *The University of Georgia Cooperative Extension*; The University of Georgia: Athens, GA, USA, 2009; p. 28.
46. Kim, J.H.; Lee, A.K.; Roh, M.S.; Suh, J.K. The effect of irradiance and temperature on the growth and flowering of Sinningia cardinalis. *Sci. Hortic. (Amst.)* **2015**, *194*, 147–153. [CrossRef]
47. World Weather Online Port Saint John's. Eastern Cape, South Africa Weather Averages | Monthly Average High and Low Temperature | Average Precipitation and Rainfall Days. Available online: https://www.worldweatheronline.com/port-saint-johns-weather-averages/eastern-cape/za.aspx (accessed on 24 March 2021).
48. Wang, X.; Zhang, W.; Miao, Y.; Gao, L. Root-Zone Warming Differently Benefits Mature and Newly Unfolded Leaves of Cucumis sativus L. Seedlings under Sub-Optimal Temperature Stress. *PLoS ONE* **2016**, *11*, e0155298. [CrossRef]
49. Morgan, J.; Moustafa, A.; Tan, A. Factors affecting the growing-on stages of lettuce and chrysanthemum in nutrient solution culture. *Acta Hortic.* **1980**, 253–262. [CrossRef]
50. Jimoh, M.O.; Afolayan, A.J.; Lewu, F.B. Heavy metal uptake and growth characteristics of Amaranthus caudatus L. under five different soils in a controlled environment. *Not. Bot. Horti Agrobot.* **2020**, *48*, 417–425. [CrossRef]
51. Steel, R.G.D.; Torrie, J.H.; Dickey, D.A. *Principles and Procedures of Statistics: A Biometrical Approach*, 3rd ed.; McGraw-Hill College: Pennsylvania, PA, USA, 1996; ISBN 978-0070610286.
52. Wahome, P.K.; Oseni, T.; Masarirambi, M. Effects of different hydroponics systems and growing media on the vegetative growth, yield and cut flower quality of gypsophila (*Gypsophila paniculata* L.). *J. Agric. Sci.* **2011**, *7*, 692–698.
53. Vogelezang, J.V.M. Effect of root-zone heating on growth, flowering and keeping quality of Saintpaulia. *Sci. Hortic. (Amst.)* **1988**, *34*, 101–113. [CrossRef]
54. Greer, D.H.; Wünsche, J.N.; Norling, C.L.; Wiggins, H.N. Root-zone temperatures affect the phenology of bud break, flower cluster development, shoot extension growth and gas exchange of "Braeburn" (*Malus domestica*) apple trees. *Tree Physiol.* **2006**, *26*, 105–111. [CrossRef] [PubMed]
55. Moss, G.I.; Dalgleish, R. Increasing returns from roses with root-zone warming. *J. Am. Soc. Hortic. Sci.* **1984**, *109*, 893–898.
56. Nxawe, S.; Laubscher, C.P.; Ndakidemi, P.A. Effect of regulated irrigation water temperature on hydroponics production of Spinach (*Spinacia oleracea* L.). *Afr. J. Agric. Res.* **2009**, *4*, 1442–1446.
57. Stoltzfus, R.M.B.; Taber, H.G.; Aiello, A.S. Effect of increasing root-zone temperature on growth and nutrient uptake by "Gold Star" muskmelon plants. *J. Plant Nutr.* **1998**, *21*, 321–328. [CrossRef]
58. Zeroni, M.; Gale, J. The effect of root temperature on the development, growth and yield of "Sonia" roses. *Sci. Hortic. (Amst.)* **1982**, *18*, 177–184. [CrossRef]
59. Gislerod, H.R. Physical conditions of propagation media and their influence on the rooting of cuttings. *Plant Soil* **1983**, *75*, 1–14. [CrossRef]
60. Andersen, C.P.; Sucoff, E.I.; Dixon, R.K. Effects of root zone temperature on root initiation and elongation in red pine seedlings. *Can. J. For. Res.* **1986**, *16*, 696–700. [CrossRef]
61. Friis, K.; Christensen, O.V. Flowering of Centradenia inaequilateralis "Cascade" as influenced by temperature and photoperiod. *Sci. Hortic. (Amst.)* **1989**, *41*, 125–130. [CrossRef]
62. Norby, R.J.; Hartz-Rubin, J.S.; Verbrugge, M.J. Phenological responses in maple to experimental atmospheric warming and CO_2 enrichment. *Glob. Chang. Biol.* **2003**, *9*, 1792–1801. [CrossRef]
63. Appelgren, M.; Heide, O.M. Regeneration in Streptocarpus Leaf Discs and its Regulation by Temperature and Growth Substances. *Physiol. Plant* **1972**, *27*, 417–423. [CrossRef]

64. van Staden, J. Changes in endogenous cytokinin levels during abscission and senescence of Streptocarpus leaves. *J. Exp. Bot.* **1973**, *24*, 667–671. [CrossRef]
65. Scott-Shaw, C.R.; Victor, J.E.; von Staden, L. *Streptocarpus formosus (Hilliard & B.L.Burtt) T.J.Edwards*; SANBI: Pretoria, South Africa, 2008.
66. Maurya, J.P.; Bhalerao, R.P. Photoperiod- and temperature-mediated control of growth cessation and dormancy in trees: A molecular perspective. *Ann. Bot.* **2017**, *120*, 351–360. [CrossRef] [PubMed]
67. Adebooye, C.O.; Schmitz-Eiberger, M.; Lankes, C.; Noga, G.J. Inhibitory effects of sub-optimal root zone temperature on leaf bioactive components, photosystem II (PS II) and minerals uptake in Trichosanthes cucumerina L. Cucurbitaceae. *Acta Physiol. Plant* **2010**, *32*, 67–73. [CrossRef]

 *horticulturae* MDPI

Communication

Evaluation of Carrageenan, Xanthan Gum and Depolymerized Chitosan Based Coatings for Pineapple Lily Plant Production

Piotr Salachna * and **Anna Pietrak**

Department of Horticulture, West Pomeranian University of Technology, 3 Papieża Pawła VI Str., 71-459 Szczecin, Poland; pa37778@zut.edu.pl
* Correspondence: piotr.salachna@zut.edu.pl; Tel.: +48-91-449-6359

Abstract: Some natural polysaccharides and their derivatives are used in horticulture to stimulate plant growth. This study investigated the effects of coating bulbs with carrageenan-depolymerized chitosan (C-DCh) or xanthan-depolymerized chitosan (X-DCh) on growth, flowering, and bulb yield as well as physiological and biochemical attributes of pineapple lily (*Eucomis autumnalis*). The results showed that treatment with C-DCh or X-DCh significantly increased all growth parameters, bulb yield, greenness index, stomatal conductance, total N, total K, and total sugar content of bulbs and accelerated anthesis as compared with untreated bulbs. The positive impact of coatings on plant growth and physiological attributes depended on the type of biopolymer complexes. The X-DCh treatment exhibited the greatest plant height, fresh weight, daughter bulb number, greenness index, stomatal conductance, total N, K, and sugar content. However, this treatment induced a significant decrease in L-ascorbic acid, total polyphenol content and antioxidant activity. Overall, the results of this study indicated high suitability of C-DCh and X-DCh as bulb coatings for pineapple lily plant production.

Keywords: biostimulants; polysaccharides; bulb coating; plant enhancement; metabolites

Citation: Salachna, P.; Pietrak, A. Evaluation of Carrageenan, Xanthan Gum and Depolymerized Chitosan Based Coatings for Pineapple Lily Plant Production. *Horticulturae* **2021**, 7, 19. https://doi.org/10.3390/horticulturae7020019

Academic Editor: Douglas D. Archbold
Received: 31 December 2020
Accepted: 27 January 2021
Published: 29 January 2021

1. Introduction

Currently, biostimulants are used to improve growth and development of horticultural plants [1,2]. Biostimulants are a broad group of substances and microorganisms with high biological activity [3–5]. Of particular interest are biodegradable polysaccharides and their depolymerized derivatives exhibiting multi-directional actions in plants [6–10]. Chitosan is one of the best known natural polysaccharides with biostimulatory properties obtained in the process of chitin de-N-acetylation [11,12]. Chitosan and its oligomeric forms have stimulated plant growth and flowering, increased photosynthesis and nutrient uptake, and protected plants against stress [13–16]. Many studies have reported the benefits provided by chitosan on various ornamental plants, such as *Begonia* × *hiemalis* Fotsch [17], *Chrysanthemum morifolium* Ramat [18], *Eucomis bicolor* Baker [19], *Freesia* × *hybrida* [20], and *Petunia* × *atkinsiana* D. Don [21]. In practice, chitosan solution is applied as a spray or drench [22–24] as well as hydrogels for coating seeds, but hydrogels from "pure" chitosan have low stability and durability [25]. Moreover, the wider use of chitosan is limited due to its poor water solubility [26]. The application of depolymerized chitosan with low molecular weight and ionic biopolymers in the form of hydrogel coating formed on the surface of plant organs based on polyelectrolyte complexes may be the solution to the problem [27]. This type of coating formed by chitosan and ionic polymers can positively affect plant growth and flowering [28]; however, it is reasonable to conduct broader research, including evaluation of the effectiveness of various biopolymers as coating components [29]. Carrageenans are a family of anionic polymers extracted from red algae used as plant biostimulants [10,30,31]. Carrageenans and their breakdown products can stimulate plant productivity and root system development, and enhance net photosynthesis, basal, and secondary metabolisms [32–36]. Among natural biopolymers, xanthan gum, an anionic, high-molecular-weight exo-polysaccharide secreted by the bacterium *Xanthomonas campestris* is also known [37]. Application of xanthan gum can influence plant growth and physiology, content of phenolic compounds, and antioxidant activity [38–40]. Xanthan gum used in micropropagation as an alternative to agar has a positive effect on the regenerative potential of some plants [41], which may indicate its biostimulative action. However, no information is available regarding the effect of xanthan gum as a biostimulant on plant growth and flowering.

Pineapple lily (*Eucomis autumnalis* (Mill.) Chitt. Asparagaceae) is a prospective bulbous ornamental plant grown in gardens, for cut flowers, and as a potted plant for indoor display [42–44]. The bulbs produce a rosette of smooth leaves and original decorative raceme-type inflorescences with a tuft of leaf-like bracts on top, composed of star-shaped white flowers with a pleasant scent. After flowering, the plants set decorative and durable green capsules. Besides its ornamental use, pineapple lily is one of the most popular plant species in traditional medicine in southern Africa [45]. The extracts of pineapple lily exert multidirectional effects, including antioxidant, anti-inflammatory, bactericidal, fungicidal and cytostatic effects [45,46]. The species is threatened with extinction in its natural habitat due to the excessive collection of bulbs for medicinal purposes as well as low vegetative propagation rate [45]. Thus, proper production methods of pineapple lily using various plant biostimulants is needed [47,48].

Previous work [49] reported that oligochitosan and sodium alginate can be successfully used for the preparation of hydrogel coatings for the bulbs of pineapple lily. However, systematic study of the effects of other biopolymers on growth, plant physiological status, and biochemical parameters of pineapple lily remains to be investigated. The current study was aimed to compare the effects of coatings containing hydrogels based on carrageenan or xanthan gum with depolymerized chitosan on the growth characteristics, flowering, bulb yield, physiological parameters, nutrients, L-ascorbic acid, and total polyphenol content, as well as antioxidant activity of pineapple lily. It was hypothesized that a coating treatment with polysaccharides would enhance the growth and bulb production of pineapple lily.

2. Materials and Methods

2.1. Plant Culture and Treatment

Bulbs of pineapple lily (*E. autumnalis*) with 12–14 cm circumference were imported from The Netherlands by Ogrodnictwo Wiśniewski Jacek Junior (Góraszka, Poland) and treated for 30 min in a suspension of 0.7% Topsin M 500 SC and 1% Captan 50 WP fungicides. Before planting, the uniform bulbs were coated according to the technology described by Startek et al. [29] in hydrogels based on 1% (*w/v*) carrageenan or 1% (*w/v*) xanthan gum in which bulbs were dipped for 30 s, and 0.2% (*w/v*) depolymerized chitosan in which the bulbs were soaked for 10 min. Control bulbs were soaked in distilled water. Depolymerized chitosan obtained by controlled free radical degradation [28] had a molecular weight of 154, 500 g mol^{-1}, the number-average molecular weight of 22,800 g mol^{-1}, and deacetylation degree of 85%. Carrageenan-depolymerized chitosan (C-DCh) and xanthan-depolymerized chitosan (X-DCh) were produced. Iota-carrageenan and xanthan gum were purchased from Sigma-Aldrich. Polysaccharides were prepared by solubilization using a magnetic stirrer. Each treatment was replicated four times and each replicate had 10 bulbs.

Coated bulbs were planted in a randomized block design on 15 April 2016 and 13 April 2017 into polyethylene boxes (60 × 40 × 19 cm) filled with peat substrates (pH 6.3) supplemented with a fertilizer Hydrocomplex (12% N, 4.5% P, and 15% K plus micronutrients; Yara International ASA, Oslo, Norway) at a dose of 3 g L^{-1}. Each box contained 10 bulbs. The boxes were transferred to a non-heated tunnel covered with a double layer of plastic located in the area of West Pomeranian University of Technology in Szczecin (53°25′ N, 14°32′ E; 25 m a.s.l.). Air temperature inside the tunnel was controlled with vents that were opened when the temperature exceeded 20 °C.

2.2. Measurement of Growth Parameters

The number of days to anthesis was recorded. When the first flowers opened in the raceme, plant height, diameter of the plant, and inflorescence length were recorded. At the end of the flowering period, the number of florets in the inflorescence were counted and fresh weight of the excised aboveground part was measured. On 3 October 2016 and 6 October 2017, the plants were removed from boxes, and fresh weight of bulbs per plant and the number of daughter bulbs were determined.

2.3. Measurement of Physiological Parameters

At the flowering stage, relative leaf chlorophyll content measurements were performed using a SPAD-502 Chlorophyll Meter (Minolta, Osaka, Japan) and stomatal conductance was assessed with a SC-1 Leaf Porometer (Dekagon Device, Pullwan, WA, USA). SPAD and stomatal conductance measurements were calculated based on four readings of four uniform leaves selected from five plants of each treatment.

2.4. Total N, P, K, and Total Sugar Content in Bulb Determination

At the end of the growing season, the bulb samples were collected, dried at 65 °C for 72 h and ground. Powdered samples (2.0 g) were digested in 17 mL concentrated 96–97% H_2SO_4. The total forms of N, P, and K were determined as outlined by Ostrowska [50]. Total N was determined according to the Kjeldahl method, P with colorimetric method according to Barton, and K by flame photometry [50]. The content of total sugar in samples of fresh bulbs was determined following the Luff-Schoorl method [51]. Nutrients and total sugar content were determined using three replicates per treatment.

2.5. L-Ascorbic Acid, Total Polyphenol Content, and Antioxidant Activity Determination

At the flowering stage, fully developed leaves were taken for biochemical analyses. Before homogenization, leaves were washed with water to remove soil, cut into slices, and dried in a circulating-air oven (35 °C ± 2 °C). Vitamin C was determined as L-ascorbic acid by the Tillman's titration method of the reduction of 2.6-dichlorophenolindophenol [52]. The preparation of plant extracts for the determination of the total polyphenol content and antioxidant activities was performed using the method of Wojdyło et al. [53] with some modifications. The sample of leaves was treated with 70% aqueous methanol (MeOH). Total polyphenol content was analyzed spectrophotometrically using the Folin–Ciocalteu colorimetric method as described by Wojdyło et al. [53]. The absorbance was measured at 760 nm. Antioxidant activity of leaves on DPPH (2,2-diphenyl-1-picrylhydrazyl) radical was determined according to the procedure of Yen and Chen [54], and DPPH inhibition percentage was calculated according to the formula provided by Rossi et al. [55]. All determinations were carried out in three replicates.

2.6. Statistical Analysis

Data were normally distributed and passed Levene's test ($\alpha \leq 0.05$) for homogeneity of variance. Data were statistically analyzed by one-way ANOVA using Statistica™ Professional 13.3.0 software (TIBCO Statistica, Palo Alto, CA, USA). After checking the goodness of fit of the model, post hoc comparisons were done using the Duncan's Multiple Range Test (DMRT) at $\alpha \leq 0.05$. The results are presented as a mean from two years of the study.

3. Results

The effect of coating with carrageenan-depolymerized chitosan (C-DCh) or xanthan-depolymerized chitosan (X-DCh) on growth and flowering of pineapple lily is shown in Table 1. The C-DCh and X-DCh applications significantly increased plant height by 8% and 16%, respectively, plant width by 39% and 40%, respectively, fresh weight of the aboveground part by 71% and 95%, respectively, inflorescence length by 32% and 25%, respectively, and the number of florets by 8% and 9%, respectively, as well as accelerated flowering by 17 and 13 days, respectively. Statistically significant differences were observed between X-DCh and C-DCh treatments. Bulbs coated in X-DCh were taller by 7% and had a greater fresh weight of the aboveground part by 16%, compared with bulbs coated in C-DCh.

Table 1. Growth and flowering parameters of pineapple lily treated with carrageenan-depolymerized chitosan (C-DCh) or xanthan-depolymerized chitosan (X-DCh).

Parameter	Type of Coatings		
	Control	C-DCh	X-DCh
Plant height (cm)	32.6 ± 0.96 c [z]	35.1 ± 0.59 b	37.7 ± 1.07 a
Plant diameter (cm)	25.7 ± 1.61 b	35.8 ± 1.86 a	36.0 ± 1.17 a
Fresh weight of the aboveground part (g)	107 ± 1.91 c	183 ± 4.10 b	209 ± 4.07 a
Length of inflorescence (cm)	16.5 ± 1.70 b	21.7 ± 1.19 a	20.7 ± 0.72 a
Number of florets	72.6 ± 3.39 b	78.6 ± 1.46 a	79.2 ± 1.15 a
Days to anthesis	161 ± 2.08 a	144 ± 3.21 b	148 ± 2.52 b

[z] Means (±SD) followed by the same small letter in the same row did not differ by Duncan's Multiple Range Test at $\alpha \leq 0.05$.

The fresh weight of bulbs, number of daughter bulbs, total N, K, and total sugar content in the pineapple lily bulbs were significantly affected by C-DCh or X-DCh complexes (Table 2). The coating of bulbs with C-DCh and X-DCh enhanced fresh weight of bulbs by 39% and 61%, respectively, and number of daughter bulbs by 24% and 48%, respectively. Moreover, the application of C-DCh and X-DCh increased levels of N by 49% and 54%, respectively, K by 46% and 57%, respectively, and total sugar content by 12% and 17%, respectively, in comparison with the control. The treatment with X-DCh resulted in the greatest fresh weight of bulbs, number of daughter bulbs, and total N, K, and sugar content. Bulb treatment with C-DCh and X-DCh did not affect total P content.

Table 2. Fresh weight of bulbs, number of daughter bulbs, total N, P, K, and total sugar content in bulb of pineapple lily treated with carrageenan-depolymerized chitosan (C-DCh) or xanthan-depolymerized chitosan (X-DCh).

Parameter	Type of Coatings		
	Control	C-DCh	X-DCh
Fresh weight of bulbs (g)	31.0 ± 1.87 c [z]	43.0 ± 2.61 b	50.0 ± 4.95 a
Number of daughter bulbs	0.75 ± 0.15 c	0.93 ± 0.07 b	1.11 ± 0.10 a
Total N content (% DW)	0.39 ± 0.03 c	0.58 ± 0.02 b	0.64 ± 0.02 a
Total P content (% DW)	0.05 ± 0.01 a	0.06 ± 0.01 a	0.05 ± 0.01 a
Total K content (% DW)	0.44 ± 0.02 c	0.64 ± 0.03 b	0.69 ± 0.02 a
Total sugar content (% FW)	6.66 ± 0.39 b	7.47 ± 0.15 a	7.76 ± 0.16 a

[z] Means (±SD) followed by the same small letter in the same row did not differ by Duncan's Multiple Range Test (DMRT) at $\alpha \leq 0.05$.

As shown in Figure 1, SPAD chlorophyll meter measurements and stomatal conductance were significantly increased due to C-DCh and X-DCh treatment in comparison to control. The C-DCh increased SPAD and stomatal conductance by 11% and 55%, and X-DCh by 7% and 31%, respectively. The SPAD and stomatal conductance of plants treated with X-DCh were 3% and 19%, respectively, greater than that of C-DCh treatment.

Figure 2 shows the effects of bulb coatings on the content of L-ascorbic acid and total polyphenols and the antioxidant activity. In comparison with the control the application of C-DCh or X-DCh significantly decreased total polyphenol content by 13% and 17%, respectively. Furthermore, the application of X-DCh significantly decreased L-ascorbic acid content by 33% and free DPPH radicals by 56%. The plant L-ascorbic acid content and antioxidant activity showed no statistically significant differences between control and C-DCh treatment.

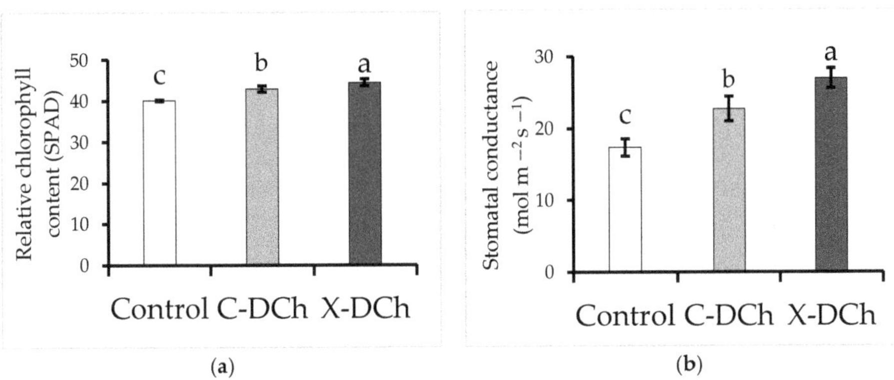

Figure 1. Relative leaf chlorophyll content (SPAD) (**a**) and stomatal conductance (**b**) of pineapple lily treated with carrageenan-depolymerized chitosan (C-DCh) and xanthan-depolymerized chitosan (X-DCh). Data are presented as means (±SD) and bars with different letters in each graph are significantly different by Duncan's Multiple Range Test (DMRT) at $\alpha \leq 0.05$.

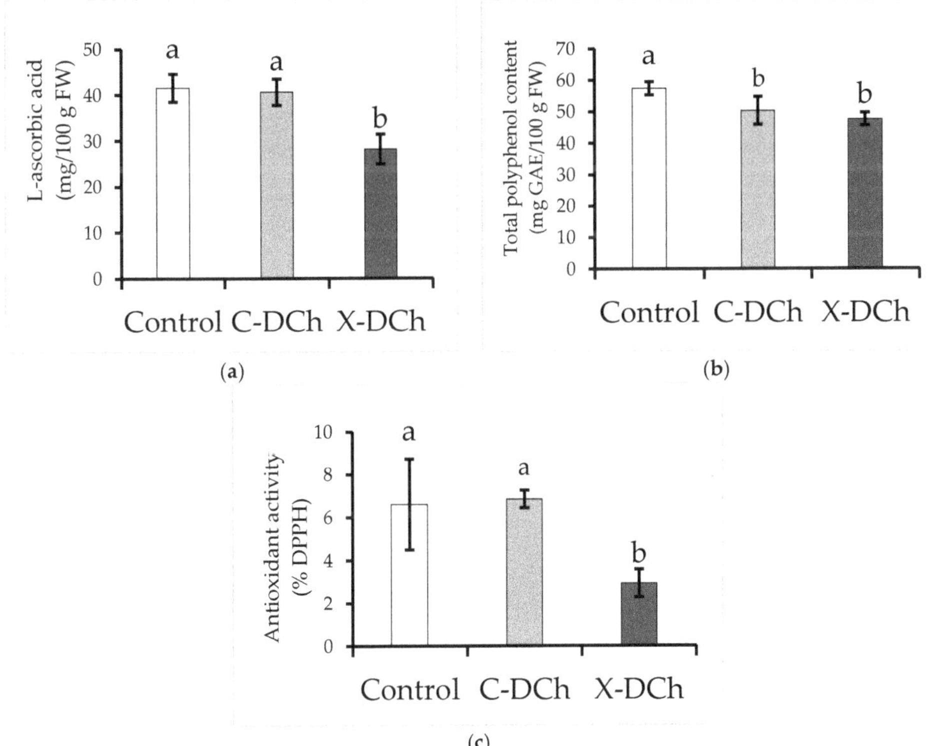

Figure 2. L-ascorbic acid (**a**) and total polyphenol content (**b**) and antioxidant activity (DPPH) (**c**) of pineapple lily treated with carrageenan-depolymerized chitosan (C-DCh) and xanthan-depolymerized chitosan (X-DCh). Data are presented as means (±SD) and bars with different letters in each graph are significantly different by Duncan's Multiple Range Test (DMRT) at $\alpha \leq 0.05$.

4. Discussion

Most studies on the application of biostimulant coatings focus on seeds, while data on coating other plant organs such as bulbs, tubers or rhizomes are far less available [56]. Our study is the first to use two biostimulant complexes, containing depolymerized chitosan and carrageenan (C-DCh) or xanthan (X-DCh), for coating pineapple lily bulbs. We found a stimulatory effect of both types of coatings on plant growth and development manifested in accelerated flowering and clearly higher yield of flowers and bulbs. In addition, plants grown out of biostimulant-coated bulbs featured more efficient gas exchange and better nutrient content. Some results of the current study are in agreement with our earlier work [29]. We reported a considerable improvement in plant growth, as assessed by morphological and physiological parameters and nutrient content in *E. autumnalis* when the bulbs were coated with oligochitosan and sodium alginate, and in *Ornithogalum saundersiae* Baker when the bulbs were coated with chitooligosaccharide and sodium alginate, carrageenan, gellan gum, or xanthan gum [28,49]. Stimulating effects of the coatings were probably due to the fact that their components enhanced plant tolerance to stresses from the beginning of their development, similar to that noted with seed coating [56]. Chitosan can improve growth and structure of a plant root system by limiting the presence of soil pathogens [11,12]. As a source of carbon for microorganisms, it also indirectly boosts soil microbial activity and thus improves absorption of minerals and water by plant roots [15]. In consequence, plants grow faster and stronger and produce better yield. Carrageenan may act as an elicitor that induces plant defense response against viroids, viruses, bacteria, or fungi, and it also may improve plant growth by controlling numerous metabolic processes including photosynthesis and assimilation of nitrogen and sulfur [10,30,35]. Xanthan gum is also capable of inducing local and systemic resistance against diseases and shows the same efficiency in plant protection against some phytopathogens as fungicides [38,57]. Soil application of xanthan gum may increase root biomass production and plant tolerance to drought and other environmental stresses [40].

The results presented in this paper demonstrated that the positive effects of coating pineapple lily bulbs on plant growth depended on the type of biopolymer complexes. The strongest plant growth stimulation was observed in plants obtained from bulbs treated with X-DCh complex. We assume that joint application of the biostimulants in X-DCh coatings may induce a stronger synergistic effect than C-DCh coating alone. It is commonly known that many biologically active substances change their properties when interacting with other substances [58,59]. Interestingly, the stimulating effect of X-DCh on the growth of the aboveground plant tissues and bulb biomass and the content of N, K, and total sugars in pineapple lily was accompanied by a clear drop in the levels of L-ascorbic acid and total polyphenols and by reduced antioxidant activity, responses not recorded in plants treated with C-DCh. The inhibitory effect of X-DCh treatment on secondary metabolite production may be a result of a trade-off between the production of plant biomass and secondary metabolism [60]. It is well known that defense and plant growth cannot usually be successfully executed at the same time [61,62]. Another possible interpretation, in line with previously cited research, is that xanthan gum induced a reduction in polyphenol content due to activation of some cellular biochemical mechanisms involved in plant resistance [57]. Still, further studies are necessary to validate either of these hypotheses.

5. Conclusions

The biostimulant complexes carrageenan-depolymerized chitosan (C-DCh) and xanthan-depolymerized chitosan (X-DCh) used for bulb coating improved plant productivity, allowing growers to speed up the production cycle in protected culture and to obtain higher quality flowers and bulbs of pineapple lily. Particularly strong biostimulant activity was shown for coatings containing derivatives of X-DCh. Bulb coatings in biostimulants seems a prospective, efficient, and environmentally friendly method of improving plant growth that can be recommended for sustainable production of ornamental plants.

Author Contributions: Conceptualization, methodology, formal analysis and investigation, P.S.; writing, P.S., A.P. All authors have read and agreed to the published version of the manuscript.

Funding: This research received no external funding.

Institutional Review Board Statement: Not applicable.

Informed Consent Statement: Not applicable.

Data Availability Statement: Data sharing not applicable.

Conflicts of Interest: The authors declare no conflict of interest.

References

1. Malik, A.; Mor, V.S.; Tokas, J.; Punia, H.; Malik, S.; Malik, K.; Sangwan, S.; Tomar, S.; Singh, P.; Singh, N.; et al. Biostimulant-Treated Seedlings under Sustainable Agriculture: A Global Perspective Facing Climate Change. *Agronomy* **2021**, *11*, 14. [CrossRef]
2. Drobek, M.; Frąc, M.; Cybulska, J. Plant Biostimulants: Importance of the Quality and Yield of Horticultural Crops and the Improvement of Plant Tolerance to Abiotic Stress—A Review. *Agronomy* **2019**, *9*, 335. [CrossRef]
3. Rouphael, Y.; Colla, G. Biostimulants in Agriculture. *Front. Plant Sci.* **2020**, *11*, 40. [CrossRef]
4. Zulfiqar, F.; Casadesús, A.; Brockman, H.; Munné-Bosch, S. An Overview of Plant-Based Natural Biostimulants for Sustainable Horticulture with a Particular Focus on Moringa Leaf Extracts. *Plant Sci.* **2020**, *295*, 110194. [CrossRef] [PubMed]
5. Kocira, A.; Lamorska, J.; Kornas, R.; Nowosad, N.; Tomaszewska, M.; Leszczyńska, D.; Kozłowicz, K.; Tabor, S. Changes in Biochemistry and Yield in Response to Biostimulants Applied in Bean (*Phaseolus vulgaris* L.). *Agronomy* **2020**, *10*, 189. [CrossRef]
6. Elarroussia, H.; Elmernissia, N.; Benhimaa, R.; El Kadmiria, I.M.; Bendaou, N.; Smouni, A.; Wahbya, I. Microalgae polysaccharides a promising plant growth biostimulant. *J. Algal Biomass Utln.* **2016**, *7*, 55–63.
7. Yang, J.; Shen, Z.; Sun, Z.; Wang, P.; Jiang, X. Growth Stimulation Activity of Alginate-Derived Oligosaccharides with Different Molecular Weights and Mannuronate/Guluronate Ratio on *Hordeum vulgare* L. *J. Plant Growth Regul.* **2020**, 1–10. [CrossRef]
8. Zhang, C.; Wang, W.; Zhao, X.; Wang, H.; Yin, H. Preparation of alginate oligosaccharides and their biological activities in plants: A review. *Carbohydr. Res.* **2020**, *494*, 108056. [CrossRef]
9. Salachna, P. Effects of Depolymerized Gellan with Different Molecular Weights on the Growth of Four Bedding Plant Species. *Agronomy* **2020**, *10*, 169. [CrossRef]
10. Shukla, P.S.; Borza, T.; Critchley, A.T.; Prithiviraj, B. Carrageenans from Red Seaweeds as Promoters of Growth and Elicitors of Defense Response in Plants. *Front. Mar. Sci.* **2016**, *3*, 81. [CrossRef]
11. Ahmed, K.B.M.; Khan, M.M.A.; Siddiqui, H.; Jahan, A. Chitosan and Its Oligosaccharides, a Promising Option for Sustainable Crop Production-a Review. *Carbohydr. Polym.* **2020**, *227*, 115331. [CrossRef] [PubMed]
12. Chakraborty, M.; Hasanuzzaman, M.; Rahman, M.; Khan, M.A.R.; Bhowmik, P.; Mahmud, N.U.; Tanveer, M.; Islam, T. Mechanism of Plant Growth Promotion and Disease Suppression by Chitosan Biopolymer. *Agriculture* **2020**, *10*, 624. [CrossRef]
13. Ohta, K.; Morishita, S.; Suda, K.; Kobayashi, N.; Hosoki, T. Effects of Chitosan Soil Mixture Treatment in the Seedling Stage on the Growth and Flowering of Several Ornamental Plants. *J. Jpn. Soc. Hortic. Sci.* **2004**, *73*, 66–68. [CrossRef]
14. Hernández-Hernández, H.; Juárez-Maldonado, A.; Benavides-Mendoza, A.; Ortega-Ortiz, H.; Cadenas-Pliego, G.; Sánchez-Aspeytia, D.; González-Morales, S. Chitosan-PVA and Copper Nanoparticles Improve Growth and Overexpress the SOD and JA Genes in Tomato Plants under Salt Stress. *Agronomy* **2018**, *8*, 175. [CrossRef]
15. Malerba, M.; Cerana, R. Chitin- and Chitosan-Based Derivatives in Plant Protection against Biotic and Abiotic Stresses and in Recovery of Contaminated Soil and Water. *Polysaccharides* **2020**, *1*, 21–30. [CrossRef]
16. ALKahtani, M.D.F.; Attia, K.A.; Hafez, Y.M.; Khan, N.; Eid, A.M.; Ali, M.A.M.; Abdelaal, K.A.A. Chlorophyll Fluorescence Parameters and Antioxidant Defense System Can Display Salt Tolerance of Salt Acclimated Sweet Pepper Plants Treated with Chitosan and Plant Growth Promoting Rhizobacteria. *Agronomy* **2020**, *10*, 1180. [CrossRef]
17. Chen, Y.E.; Yuan, S.; Liu, H.M.; Chen, Z.Y.; Zhang, Y.H.; Zhang, H.Y. A combination of chitosan and chemical fertilizers improves growth and disease resistance in *Begonia* × *hiemalis* Fotsch. *Hortic. Environ. Biotechnol.* **2016**, *57*, 1–10. [CrossRef]
18. Elansary, H.O.; Abdel-Hamid, A.M.; Yessoufou, K.; Al-Mana, F.A.; El-Ansary, D.O.; Mahmoud, E.A.; Al-Yafrasi, M.A. Physiological and molecular characterization of water-stressed Chrysanthemum under robinin and chitosan treatment. *Acta Physiol. Plant.* **2020**, *42*, 31. [CrossRef]
19. Byczyńska, A. Chitosan improves growth and bulb yield of pineapple lily (*Eucomis bicolor* Baker) an ornamental and medicinal plant. *World Sci. News* **2018**, *110*, 159–171.
20. Salachna, P.; Zawadzińska, A. Effect of chitosan on plant growth, flowering and corms yield of potted freesia. *J. Ecol. Eng.* **2014**, *15*, 97–102.
21. Krupa-Małkiewicz, M.; Fornal, N. Application of chitosan in vitro to minimize the adverse effects of salinity in *Petunia* × *atkinsiana* D. *Don. J. Ecol. Eng.* **2018**, *19*, 143–149. [CrossRef]
22. Malerba, M.; Cerana, R. Recent advances of chitosan applications in plants. *Polymers* **2018**, *10*, 118. [CrossRef] [PubMed]
23. El-Miniawy, S.M.; Ragab, M.E.; Youssef, S.M.; Metwally, A.A. Response of strawberry plants to foliar spraying of chitosan. *Res. J. Agric. Biol. Sci.* **2013**, *9*, 366–372.

24. Algam, S.A.E.; Xie, G.; Li, B.; Yu, S.; Su, T.; Larsen, J. Effects of Paenibacillus Strains and Chitosan on Plant Growth Promotion and Control of Ralstonia Wilt in Tomato. *J. Plant Pathol.* **2010**, *92*, 593–600.
25. Muhmed, S.A.; Nor, N.A.M.; Jaafar, J.; Ismail, A.F.; Othman, M.H.D.; Rahman, M.A.; Aziz, F.; Yusof, N. Emerging Chitosan and Cellulose Green Materials for Ion Exchange Membrane Fuel Cell: A Review. *Energy Ecol. Environ.* **2020**, *5*, 85–107. [CrossRef]
26. Szymańska, E.; Winnicka, K. Stability of chitosan—a challenge for pharmaceutical and biomedical applications. *Mar. Drugs* **2015**, *13*, 1819–1846. [CrossRef]
27. Bartkowiak, A.; Startek, L.; Salachna, P.; Zurawik, P. Method of Hydrogel Coating Formation on the Surface of Plant Organs. *Pat. No PL* **2008**, *197101*, 29.
28. Salachna, P.; Grzeszczuk, M.; Soból, M. Effects of Chitooligosaccharide Coating Combined with Selected Ionic Polymers on the Stimulation of *Ornithogalum saundersiae* Growth. *Molecules* **2017**, *22*, 1903. [CrossRef]
29. Startek, L.; Bartkowiak, A.; Salachna, P.; Kaminska, M.; Mazurkiewicz-Zapalowicz, K. The influence of new methods of corm coating on freesia growth, development and health. *Acta Hortic.* **2005**, *673*, 611–616. [CrossRef]
30. Vera, J.; Castro, J.; Contreras, R.A.; González, A.; Moenne, A. Oligo-carrageenans induce a long-term and broad-range protection against pathogens in tobacco plants (Var. Xanthi). *Physiol. Mol. Plant Pathol.* **2012**, *79*, 31–39. [CrossRef]
31. Abad, L.V.; Aurigue, F.B.; Relleve, L.S.; Montefalcon, D.R.V.; Lopez, G.E.P. Characterization of low molecular weight fragments from gamma irradiated κ-carrageenan used as plant growth promoter. *Radiat. Phys. Chem.* **2016**, *118*, 75–80. [CrossRef]
32. Vanegas, J.S.; Torres, G.R.; Campos, B.B. Characterization of a κ-Carrageenan Hydrogel and Its Evaluation as a Coating Material for Fertilizers. *J. Polym. Environ.* **2019**, *27*, 774–783. [CrossRef]
33. Ahmad, B.; Jahan, A.; Sadiq, Y.; Shabbir, A.; Jaleel, H.; Khan, M.M.A. Radiation-Mediated Molecular Weight Reduction and Structural Modification in Carrageenan Potentiates Improved Photosynthesis and Secondary Metabolism in Peppermint (*Mentha piperita* L.). *Int. J. Biol. Macromol.* **2019**, *124*, 1069–1079. [CrossRef] [PubMed]
34. Akalin, G.O.; Pulat, M. Controlled Release Behavior of Zinc-Loaded Carboxymethyl Cellulose and Carrageenan Hydrogels and Their Effects on Wheatgrass Growth. *J. Polym. Res.* **2020**, *27*, 1–11. [CrossRef]
35. Saucedo, S.; González, A.; Gómez, M.; Contreras, R.A.; Laporte, D.; Sáez, C.A.; Zúñiga, G.; Moenne, A. Oligo-Carrageenan Kappa Increases Glucose, Trehalose and TOR-P and Subsequently Stimulates the Expression of Genes Involved in Photosynthesis, and Basal and Secondary Metabolisms in *Eucalyptus globulus*. *BMC Plant Biol.* **2019**, *19*, 1–13. [CrossRef]
36. San, P.T.; Khanh, C.M.; Khanh, H.H.N.; Khoa, T.A.; Hoang, N.; Nhung, L.T.; Trinh, N.T.K.; Nguyen, T.-D. k-Oligocarrageenan Promoting Growth of Hybrid Maize: Influence of Molecular Weight. *Molecules* **2020**, *25*, 3825. [CrossRef]
37. Kumar, A.; Rao, K.M.; Han, S.S. Application of xanthan gum as polysaccharide in tissue engineering: A review. *Carbohydr. Polym.* **2018**, *180*, 128–144. [CrossRef]
38. Luiz, C.; Schauffler, G.P.; Lemos-Blainski, J.M.; Rosa, D.J.; Di Piero, R.M. Mechanisms of action of aloe polysaccharides and xanthan gum for control of black rot in cauliflower. *Sci. Hortic.* **2016**, *200*, 170–177. [CrossRef]
39. Van Binh, N.; Diep, T.B.; Sang, H.D.; Thao, H.P.; Thom, N.T.; Quynh, T.M. Low Molecular Weight Xanthan Prepared by Gamma Irradiation and Its Effects on Development of Seedlings. *RAD Conf. Proc.* **2016**, *1*, 95–98. [CrossRef]
40. Jeong, H.; Jang, H.Y.; Ahn, S.J.; Kim, E. β-Glucan-and Xanthan Gum-Based Biopolymer Stimulated the Growth of Dominant Plant Species in the Korean Riverbanks. *Ecol. Resilient Infrastruct.* **2019**, *6*, 163–170.
41. Jain, R.; Babbar, S.B. Xanthan gum: An economical substitute for agar in plant tissue culture media. *Plant Cell Rep.* **2006**, *25*, 81–84. [CrossRef] [PubMed]
42. Carlson, A.S.; Dole, J.M. Postharvest handling recommendations for cut pineapple lily. *HortTechnology* **2014**, *24*, 731–735. [CrossRef]
43. Carlson, A.S.; Dole, J.M. Determining Optimal Bulb Storage and Production Methods for Successful Forcing of Cut Pineapple Lily. *HortTechnology* **2015**, *25*, 608–616. [CrossRef]
44. Salachna, P.; Grzeszczuk, M.; Meller, E.; Soból, M. Oligo-Alginate with Low Molecular Mass Improves Growth and Physiological Activity of *Eucomis autumnalis* under Salinity Stress. *Molecules* **2018**, *23*, 812. [CrossRef]
45. Masondo, N.A.; Finnie, J.F.; Van Staden, J. Pharmacological potential and conservation prospect of the genus *Eucomis* (Hyacinthaceae) endemic to southern Africa. *J. Ethnopharmacol.* **2014**, *151*, 44–53. [CrossRef]
46. Salachna, P.; Grzeszczuk, M.; Wilas, J. Total phenolic content, photosynthetic pigment concentration and antioxidant activity of leaves and bulbs of selected *Eucomis* L'Hér. taxa. *Fresen. Environ. Bull.* **2015**, *24*, 4220–4225.
47. Salachna, P.; Mizielińska, M.; Soból, M. Exopolysaccharide Gellan Gum and Derived Oligo-Gellan Enhance Growth and Antimicrobial Activity in Eucomis Plants. *Polymers* **2018**, *10*, 242. [CrossRef]
48. Salachna, P.; Zawadzińska, A. Effect of nitric oxide on growth, flowering and bulb yield of *Eucomis autumnalis*. *Acta Hortic.* **2017**, *1201*, 635–640.
49. Salachna, P.; Zawadzińska, A. Comparison of morphological traits and mineral content in *Eucomis autumnalis* (Mill.) Chitt. plants obtained from bulbs treated with fungicides and coated with natural polysaccharides. *J. Ecol. Eng.* **2015**, *16*, 126–142. [CrossRef]
50. Ostrowska, A.; Gawliński, S.; Szczubiałka, Z. *Methods for Analyzing and Assessing the Properties of Soil and Plants*; Instytut Ochrony Środowiska: Warsaw, Poland, 1991; pp. 1–333.
51. Salachna, P.; Mikiciuk, M.; Zawadzińska, A.; Piechocki, R.; Ptak, P.; Mikiciuk, G.; Pietrak, A.; Łopusiewicz, Ł. Changes in Growth and Physiological Parameters of ×*Amarine* Following an Exogenous Application of Gibberellic Acid and Methyl Jasmonate. *Agronomy* **2020**, *10*, 980. [CrossRef]

52. *AOAC Official Methods of Analysis of the Association of Official Analytical Chemists*; Association of Official Analytical Chemists: Washington, DC, USA, 1990.
53. Wojdyło, A.; Oszmiański, J.; Czemerys, R. Antioxidant activity and phenolic compounds in 32 selected herbs. *Food Chem.* **2007**, *105*, 940–949. [CrossRef]
54. Yen, G.C.; Chen, H.Y. Antioxidant activity of various tea extracts in relation to their antimutagenicity. *J. Agric. Food Chem.* **1995**, *43*, 27–32. [CrossRef]
55. Rossi, M.; Giussani, E.; Morelli, R.; Scalzo, R.; Nani, R.C.; Torreggiani, D. Effect of fruit blanching on phenolics and radical scavenging activity of highbush blueberry juice. *Food Res. Int.* **2003**, *36*, 999–1005. [CrossRef]
56. Ma, Y. Seed coating with beneficial microorganisms for precision agriculture. *Biotechnol. Adv.* **2019**, *37*, 107423. [CrossRef] [PubMed]
57. Antoniazzi, N.; Deschamps, C.; Bach, E.E. Effect of xanthan gum and allicin as elicitors against Bipolaris sorokiniana on barley in field experiments. *J. Plant Dis. Prot.* **2008**, *115*, 104–107. [CrossRef]
58. Rademacher, W. Plant growth regulators: Backgrounds and uses in plant production. *J. Plant Growth Regul.* **2015**, *34*, 845–872. [CrossRef]
59. Rouphael, Y.; Colla, G. Synergistic biostimulatory action: Designing the next generation of plant biostimulants for sustainable agriculture. *Front. Plant Sci.* **2018**, *9*, 1655. [CrossRef]
60. Bot, J.L.; Benard, C.; Robin, C.; Bourgaud, F.; Adamowicz, S. The 'trade-off' between synthesis of primary and secondary compounds in young tomato leaves is altered by nitrate nutrition: Experimental evidence and model consistency. *J. Exp. Bot.* **2009**, *60*, 4301–4314. [CrossRef]
61. Caretto, S.; Linsalata, V.; Colella, G.; Mita, G.; Lattanzio, V. Carbon Fluxes between Primary Metabolism and Phenolic Pathway in Plant Tissues under Stress. *Int. J. Mol. Sci.* **2015**, *16*, 26378–26394. [CrossRef]
62. Ibrahim, M.H.; Jaafar, H.Z.E.; Rahmat, A.; Rahman, Z.A. Effects of Nitrogen Fertilization on Synthesis of Primary and Secondary Metabolites in Three Varieties of Kacip Fatimah (*Labisia pumila* Blume). *Int. J. Mol. Sci.* **2011**, *12*, 5238–5254. [CrossRef]

Article

Environmental Analysis of Sustainable Production Practices Applied to Cyclamen and Zonal Geranium

Jaco Emanuele Bonaguro, Lucia Coletto, Paolo Sambo, Carlo Nicoletto and Giampaolo Zanin *

Department of Agronomy, Food, Natural Resources, Animals and Environment (DAFNAE), University of Padova, Viale dell'Università, 16, 35020 Legnaro, Italy; jaco.bonaguro@gmail.com (J.E.B.); colettolucia@gmail.com (L.C.); paolo.sambo@unipd.it (P.S.); carlo.nicoletto@unipd.it (C.N.)
* Correspondence: paolo.zanin@unipd.it; Tel.: +39-049-827-2902

Abstract: Italian floriculture is facing structural changes. Possible options to maintain competitiveness of the involved companies include promotion of added values, from local production to environmental sustainability. To quantify value and benefits of cleaner production processes and choices, a holistic view is necessary and could be provided by life cycle assessment (LCA) methodology. Previous studies on ornamental products generally focused on data from one company or a small sample. The aim of this study was a gate-to-gate life cycle assessment of two ornamental species, cyclamen (*Cyclamen persicum* Mill.) and zonal geranium (*Pelargonium* × *hortorum* Bailey), using data from a sample of 20 companies belonging to a floriculture district in the Treviso, Veneto region. We also assessed the potential benefits of the environmental impact of alternative management choices regarding plant protection and reuse of composted waste biomass. Life cycle impact assessment showed higher impact scores for the zonal geranium, mainly as a consequence of greenhouse heating with fossil fuels. This factor, along with higher uniformity of production practices and technological levels of equipment, translated to a lower variability in comparison with cyclamen production, which showed a wider results range, in particular for eutrophication, acidification and human toxicity potential. The application of integrated pest management with cyclamen had significant benefits by reducing acidification and human toxicity, while reducing use of mineral nutrients through amending growing media with compost resulted in a reduction in eutrophication potential. Similar achievable benefits for zonal geranium were not observed because of the dominant contribution of energy inputs.

Keywords: life cycle impact assessment (LCIA); plant protection; compost; sustainable greenhouse production

Citation: Bonaguro, J.E.; Coletto, L.; Sambo, P.; Nicoletto, C.; Zanin, G. Environmental Analysis of Sustainable Production Practices Applied to Cyclamen and Zonal Geranium. *Horticulturae* **2021**, *7*, 8. https://doi.org/10.3390/horticulturae7010008

Received: 20 December 2020
Accepted: 14 January 2021
Published: 15 January 2021

1. Introduction

Ornamental plant production is a specialized and intensive agricultural sector that includes a wide range of outputs, such as cut flowers, nursery stock, potted flowering or leafy plants, bulbs and tubers. Europe is the largest consumer market, with Germany, the United Kingdom, France and Italy as leading consumers. Italy is also an important producer, having over 14,000 companies with a GSP of over 1125 million € [1]. This sector has a complex structure, with a few regions having districts specialized in some sections of the production chain.

The Veneto region of Italy is home to some important districts, located in Padova, Treviso, Vicenza and Rovigo provinces. Data from 2016 [2] showed a total of 1490 companies, with a total area of 2730 ha, and a GSP of 206 million €. The overall trend compared to the previous five years highlights how the sector is facing structural changes to cope with the ongoing stagnation in domestic demand as a consequence of the current economic crisis; the number of companies is steadily decreasing, averaging −2% per year, with 4–5% peak losses in some districts (Rovigo, Vicenza). The GSP value of marketable pot plants decreased slightly (−0.5%) until 2016, then the trend started to increase. The nursery production of ornamental, vegetable and orchard plants is stable. Regarding marketing areas, local and regional sales fell (from 34.3% to 29.1% and from 22.6% to 20.2% over five years, respectively), but a slight increasing share of sales to other Italian regions (+3.9% in five years) or EU countries (+4.7% in five years). Indeed, an increased number of companies have obtained the Certificate of Conformity required for sales in EU member countries, to 264 (+18%) in 2018. The changes highlighted by these data are partly related to increased competitiveness from emerging countries [3], and partly to the shift in consumer preferences. Italian companies operating in the northern, high-cost regions are generally small and family-run, and cannot tackle sudden changes in international markets with cost reductions and technological improvements only. Possible options for maintaining competitiveness could focus on the promotion of added value, such as typical/local productions, range and variety of choice, seasonal products and "eco-friendly" choices in production systems. For countries within the EU, sustainable or cleaner production are becoming a requirement rather than an encouraged practice, even the agricultural sector which is often regarded as a polluting activity [4].

Cleaner production is defined by the United Nations Environmental Program [5] as the continuous application of an integrated preventive environmental strategy to processes, products and services, to increase overall efficiency and reduce risks to humans and the environment. Five main components of cleaner production are related to conservation of raw materials, water and energy, eliminating toxic and dangerous emissions and reducing waste. Plastic waste, fertilizer use, peat-based growing media and heating requirements are usually perceived as major contributors to protected crop impacts on environment. Some of the above-mentioned issues have been addressed by researchers, such as integrated or biological crop protection [6–9], use of slow-release fertilizers [10–13], irrigation plans based on crop needs [14], and cultivation of native low energy demanding species [15]. Use of alternative containers such as biodegradable pots for the cultivation of ornamental pot plants were evaluated in various studies [16,17]; also peat substitution with composted materials, or other agro-industrial by-products rich in nutrients, has been widely evaluated in several trials with container grown plants, such as shrubs [18], poinsettia [17], geranium [19–21] and other bedding plants [22–24]. Efficient use of energy in greenhouses has received great focus [25–27]. Many trials aimed at reducing the energy consumption of greenhouses have focused on ventilation processes and the effects of thermal energy and mass transfer [28–31].

To quantify the potential impacts and assess the efficiency of reduction measures on specific crops and production systems, a life cycle methodology should be used. Life cycle assessment (LCA) is a material and energy balance applied to the production of goods or services (ISO 14040 [32]). This methodology has been applied to some ornamental commodities and production systems [33–36]. Previous studies on potted plants under protected cultivation highlighted some of the processes and materials involved in the production of certain emissions, such as energy for heating and artificial lighting, greenhouse frames and covers, plastic containers and peat [35,37–40]. Most assessments, except for a study on nursery production conducted by Lazzerini et al. [39], analyzed data sourced from one representative company and from specific literature or databases.

Objective of the Study

The aim of this study was to assess the environmental aspects of the cultivation of two ornamental species, using data from a sample of nurseries in the Treviso production district. While trying to define average impact results for the most important categories, we analyzed how different management choices and production practices affect final results. In the following sections we present the functional units, data collection processes and the alternative scenarios we chose to assess.

2. Materials and Methods

2.1. Goal and Scope

The goal of this research was to characterize the final cultivation phase of two potted flowering plant species, cyclamen and zonal geranium, from an environmental point of view, defining a range of results representative of the most common practices in the investigated floriculture district. We also assessed the potential environmental benefits achievable with specific practices or management choices that have been adopted by individual growers independently.

Other practices that apply to all the investigated companies, such as collection and recycling of plastic materials, have been implemented in the system models. The scenarios that we investigated concern typical environmental bottlenecks of protected cultivation, such as fertilizer use, plant protection and waste management (biomass).

The scope of our study included production, installation, use and disposal of capital goods (greenhouse frame and cover, as well as heating systems and auxiliary equipment for fertigation) and production, transport, use and disposal of crop inputs. The model system we describe was based on the production practices of a sample of nurseries sited in Treviso province. Since our goal was to describe and assess common practices and average structure and technology, comparison of different company sizes or sale types were outside the scope of this study.

We used open LCA software version 1.5.0 and EcoInvent database version 3.3 to input and model the Life Cycle Inventory data. Life cycle impact assessment (LCIA) was performed with CML-2015 method (CML-Centrum Voor Milieukunde Der Rijksuniversiteit Leiden), first created by the University of Leiden in the Netherlands in 2001. It has been published in a handbook with several authors [41]. The impacts were described using the baseline method impact categories. Acidification potential (AP) measures the increase of the acidity in water and soil systems due to the acidifying effects of anthropogenic emissions nitrogen oxides (NOx) and sulfur oxides (SOx). Acidification potential is expressed using the reference unit, kg SO_2 equivalents. Global warming potential (GWP) measures the alteration of global temperature caused by greenhouse gases released by human activities; characterization of the model was based on factors developed by the UN's Intergovernmental Panel on Climate Change (IPCC). Factors are expressed as global warming potential over the time horizon of 100 years (GWP100), measured in the reference unit kg CO_2 equivalents. Eutrophication is the build-up of a concentration of chemical nutrients in an ecosystem which leads to abnormal productivity. Emissions of ammonia, nitrates, nitrogen oxides and phosphorous to air or water all have an impact on eutrophication. This category is expressed using the reference unit, kg $PO_4{}^{3-}$ equivalents. Direct and indirect impacts of fertilizers are included in the method. The direct impacts are from production of the fertilizers and the indirect ones are calculated using the IPCC method to estimate emissions to water causing eutrophication. Environmental toxicity is measured as two separate impact categories which examine freshwater and terrestrial emission of some substances, such as heavy metals, which can have impacts on the ecosystem. Assessment of toxicity is based on maximum tolerable concentrations in water for ecosystems. The calculation method provides a description of fate, exposure and the effects of toxic substances on the environment. Characterization factors are expressed using the reference unit, kg 1,4-dichlorobenzene equivalents (1,4-DB). The human toxicity potential is a calculated index that reflects the potential harm of a unit of chemical released into the environment, and it is based on both the inherent toxicity of a compound and its potential dose. This impact category is measured in 1,4-DB equivalents. The normalization step is necessary to analyze and describe the relevance of single contributions to an impact category, and to calculate the order of magnitude of the category indicator results relative to a reference information (i.e., total impacts for the selected category in a specific area). LCIA results were normalized with factors for the EU 25 area.

2.2. Data Collection

Data were collected through a survey conducted through questionnaires and interviews with 20 floriculture companies belonging to the Florveneto association, representing ornamental plant growers in Treviso province. The questionnaires were administered in person to the owners, at the company, so that the data collected could be, at least in part, verified. A questionnaire, which had previously been submitted to and validated by two pilot companies, was used to collect information on general production practices, greenhouse structures and equipment. The questionnaire with examples of compilation is available as Supplementary Material questionnaire.

Functional Units and System Boundaries

The functional unit was a single marketable plant in a 14-cm pot. The investigated species, zonal geranium and cyclamen, were chosen for several reasons: First, their economic relevance (they comprise 20% and 22% of the Italian flower market, respectively); second, they represent part of an ideal crop sequence for the average nursery. Lastly, given the seasonality of their production cycles, they are crops with different climate control needs and energy demands. System boundaries include all operations and inputs from transplant to market-ready flowering plants. The plug production phase was also included, even if specific information on seedling or cutting production for the considered species were not collected. This is also motivated by considerable differences concerning the choices of variety and young plant producers found among the surveyed companies.

2.3. System Description: Cyclamen

Cyclamen (*Cyclamen persicum* Mill.) plants are usually grown in structures with a plastic cover (single layer) over a galvanized steel frame. Average plastic cover replacement rate is 6 years, while supporting structure lifetime is 30 years. Potting of young plants occurs from May to mid-July. With an average growing period of 14–16 weeks, early potted plants bloom in September. Optimal temperature in the first period of growth is around 18–20 °C. During flower development normal temperatures should be between 15 and 20 °C. To promote cooler temperatures, shading from 30% to 50% is applied in summer months, together with lateral and roof ventilation. Active cooling systems, like fogging or fan-and-pad are installed and operating in only three nurseries. Cyclamen seedlings are transplanted into 14-cm pots, filled with a substrate composed of white peat with a coarse, porous texture (40% v/v), black peat (45–50% v/v) and expanded perlite (10–15% v/v). Plants are irrigated using overhead spray irrigation (no added fertilizer) for 1–2 weeks, then a fertilizer solution (N:P:K at 1:0.4:1.2) is applied. In some cases, overhead spray irrigation is still preferred at this stage, while most growers (14 out of 20) start fertigation with a spaghetti tube system. Fertilizer solutions applied during the growing period have increasing ratios of potassium to phosphorus to promote flowering and plant resistance to both disease and environmental stress (typical formulations: 17N-3.05P-14.2K; 20N-8.29P–23.3K). Plants are spaced after one month to allow air circulation and canopy growth. Fungal diseases include *Botrytis* and *Fusarium*, anthracnose and powdery mildew. Most are limited by prevention practices and improved breeding, yet between one and three fungicide treatments (classes: Carbamate, thiadiazole, amide, aromatic organic compounds) are reported by most growers. Common cyclamen pests are thrips (*Frankliniella occidentalis; Echinotrips americanus*), aphids (*Aphys gossypii, Aulacortum circumflexum*), vine weevil (*Otiorhynchus sulcatus*) and mites (*Steneotarsonemus pallidus, Tetranichus urticae*). Insecticides (active ingredient classes: Neonicotinoids, organophosphate, pyrethroids or avermectine) are applied from 2 to 5 times during the growing cycle. Growth regulators (chlormequat or daminozide) are applied once or twice to inhibit petiole elongation by 14 growers. See Supplementary Material questionnaire for the complete list of the inputs that were considered.

2.4. System Description: Zonal Geranium

Zonal geranium (*Pelargonium* × *hortorum* Bailey) plants are usually grown in structures with a plastic cover (double layer, air inflated) and galvanized steel frame, or in glasshouses with a steel frame. Average replacement rate of a plastic cover is 6 years while glass and supporting structures lifetime often exceeds 30 years which was the value assumed for calculations. The most widely used heating system consists of diesel-powered fan-burners generating hot air while only two companies use gas boilers and a network of polypropylene pipes to deliver hot water under cultivation benches. In the first 10–15 days after seedling transplant, optimal temperature is around 18 °C in the daytime and 16 °C at night. After this phase, diurnal temperatures are kept around 16 °C and night temperatures around 14 °C. No artificial lighting is applied during this growth phase. Growing media are usually comprised of peat moss (80–85% v/v) blended with porous materials such as perlite or expanded clay (10–15% v/v). Plants are fertigated using overhead spray irrigation for a period ranging from 6 days to 3 weeks, depending on the individual choices made by growers. After this period, until marketable size is attained, plants are placed on benches and fertigated with ebb-and-flow or with spaghetti-tubing irrigation systems. Fertilizer solutions applied during the first period have a N:P:K ratio of 1:0.5:1. To promote flower quality, potassium concentration is increased during the final growth phase (N:P:K at 0.8:0.3:1.2). Common diseases are *Xanthomonas campestris* pv. *pelargonii* (wilt and spots), *Ralstonia* (wilt), *Pythium*, and *Botrytis*. Bacterial diseases are best fought with prevention practices and early detection, and soil-borne fungal diseases can be prevented by avoiding excessive air and substrate humidity, facilitating canopy air movement and raising night temperatures. Besides prevention practices, plants are usually treated one to three times with fungicides (active ingredient classes: Dichlorophenyl dicarboximide, aromatic organic compounds, amide). As a typical spring crop, zonal geranium is very sensitive to thrips; aphids (*Acyrthosiphon malvae*) can also be a problem and cause small, distorted leaves and black sooty mold. Insecticides are applied preventively in 40% of cases; most common active ingredients belong to the carbamate, organochlorine and pyrethroid classes. Along with other ornamentals such as petunias (*Petunia* spp.) and calibrachoas (*Calibrachoa* spp.), pelargoniums can be affected by budworms (*Geraniums bronze*, *Cacyreus marshalli*) during the last growth stages. These worms can devastate geraniums by tunneling into young buds and destroying the flower. Neonicotenoid or pyrethroid insecticides are applied to control this pest. See Supplementary Material questionnaire for the complete list of the inputs that were considered.

2.5. Assumptions

Data for background processes such as material manufacturing and disposal activities were sourced from the Ecoinvent 3.3 database, and modeled with OpenLCA ver. 1.5.0. Direct emissions were calculated by using estimation models, which are flexible and allow for an estimation of mitigating options. For fertilizer use, we estimated nitrate (NO_3^-) emissions with the Swiss agricultural life cycle assessment (SALCA) method, assuming a draining fraction of 25% for open-loop systems, which is a common leaching value applied to prevent root zone salinization. Phosphate (PO_4^{3-}) emissions were calculated according to SALCA-P emission model [42]. Plant protection products applied were modeled as emissions to agricultural soil.

2.6. Description of Alternative Practices

As mentioned earlier, during the data collection it was noticed that, even if close similarities were recorded in most of the interviewed nurseries regarding structure types, technological level of growing equipment, management decisions and cultivation inputs for the studied crops, the choices made by some growers led to significant differences in the reported input levels. Management decisions could in turn lead to different emission patterns and levels. These practices mainly included plant protection practices, fertigation management and recycling of waste biomass.

2.6.1. Integrated Pest Management and Biological Plant Protection

Monitoring of insect presence (with chromotropic traps or visual inspection) is a known, yet not very widespread practice. Objective assessment of infestation and potential damage is also very difficult for crops with aesthetic value as their main feature. Despite this, the application of integrated pest management (IPM) and biological control agents is receiving growing attention, also because many active ingredients registered for use on ornamental species have recently been revoked or are no more available [43].

Due to the greater effort required, and uncertainties linked to these practices, most growers are delaying their application and still rely heavily on chemical control.

Based on information from four growers using IPM strategies, we assessed the potential impact of less chemical input and use (manufacturing of raw materials and soil emissions) as compared to an average production scenario. For cyclamen production, we considered that prevention practices at the transplant phase, with inoculation of a biological antagonist to *F. oxysporum* in the growth medium, can the reduce need for fungicide treatment to 1 per crop cycle, while improved insect scouting and monitoring reduces insecticide sprays to 2 per crop cycle. Zonal geranium benefits from biological prevention and control both at the transplant phase and in the first stages with *T. harzianum* and *B. subtilis* strains, while preventive insecticide spraying is integrated with antifeedant treatments (Azadirachtin); for this scenario we considered no fungicide treatment and a reduction of 40% in insecticide use (active substance).

2.6.2. Management and Reuse of Waste Biomass

Protected soilless crops generate a significant amount of waste, due to material requirements for growing media, containers, benches, irrigation pipes and plug trays. These materials need to be disposed of at their end-of-life and several options are available from incineration to landfilling, or composting, depending on material segregation practices, regulations and grower's choices. Recycling of plastic material is a common and well-established practice among the interviewed growers, thanks to good awareness and coordinated efforts by the Florveneto association. Management of biowaste differs between growers. The amount of non-yield biomass in ornamental containerized crops is lower than in other protected crops, yet a certain amount of unsold or discarded plants are produced and must be disposed of confined windrow composting and reuse in situ could be an option, and one grower reported to have adopted this practice. However, in this case chemical and physical properties as well as direct emissions are probably highly variable and difficult to measure, and we therefore chose to model an alternative option, where compost is produced from miscellaneous green waste in a composting facility and used in growth media preparation as a substitute for peat. The considered rate of compost addition to the growth medium is 20% (v/v); this was chosen in accordance with growth trials of containerized plants on compost amended substrates reported in several studies [19,44,45]. For different plant species, supporting effects on growth with compost rates up to 20% were reported in these studies, but different effects were found for higher substitution rates. An analysis from a local composting plant shows the chemical composition and nutrient content of composted garden waste (Table 1). Two options are considered for the offset of mineral fertilizers: NPK content of compost does not replace fertilizers (option 1); NPK content replaces part of the mineral fertilizers applied through fertigation (option 2). The following rates of nutrient content available for the crop were considered: 20% for N, 50% for P and 50% for K. These values were taken from Boldrin et al. [46], and were reduced to account for the limited length of the growing period for the considered species.

Table 1. Chemical and physical properties of the garden waste compost considered for the evaluation of the impacts.

Compost Characteristic	Value
Bulk density (kg m^{-3} dry weight)	404
Water holding capacity (v/v)	64.8
Dry matter (%)	66.5
Organic matter (%)	38.7
pH	8.70
Electrical conductivity (Sd m^{-1})	3.78
NO_3^- (mg L^{-1})	108
PO_4^{3-} (mg L^{-1})	40.6
Na^+ (mg L^{-1})	200
NH_4^+ (mg L^{-1})	19.7
K^+ (mg L^{-1})	603
Mg^{2+} (mg L^{-1})	22.8
Ca^{2+} (mg L^{-1})	115

3. Results and Discussion

The considered inputs (Supplementary Material questionnaire) were grouped into six main categories, which include production, use and end-of-life phases: Greenhouse structures and covering materials, fertilizers, plant protection products, pots (including only the containers used in the cultivation phase), growing media and heating. Looking at absolute values (Table 2) for the assessed impact categories, we can note how the heated crop (zonal geranium) scored higher results for all indicators, even by several orders of magnitude for AP and GWP categories. As highlighted in the analysis of relative contributions (Figures 1 and 2), heating with fossil fuels contributed the greatest to the inputs of production. This factor, together with the greater uniformity found for some management choices in zonal geranium, also influenced the variability, which showed minor fluctuations around average values compared to cyclamen. For geranium, to better highlight the contribution of the main groups of input to the different impact categories we excluded the most impacting one (heat) even if this is beyond both the scope of the work and the meaning of the LCA analysis. With the exclusion of heating, the categories with the greatest weight were the greenhouse structure and coverage that contributed over 60% in the fresh water aquatic ecotoxicity (FWAE), human toxicity (HT) and terrestrial ecotoxicity (TE) categories, the pot that contributed over 51% in the acidification potential (AP) and 64% in the GWP categories, the substrate accounted for about 20% in three categories (AP, GWP and HT), and the fertilizers which contributed 75% in the Eutrophication potential (EP) category.

Table 2. Absolute values and standard deviation (in percentage) for the assessed impact categories for flowering potted plants of cyclamen (*Cyclamen persicum* Mill.) and zonal geranium (*Pelargonium* × *hortorum* Bailey).

Impact Category	Reference Unit	Cyclamen		Zonal Geranium	
	Equivalents	Mean	St.Dev. (%)	Mean	St.Dev. (%)
Acidification potential (AP)	kg SO_2	0.00036	11.63	0.00175	1.28
Global warming potential (GWP)	kg CO_2	0.07459	3.32	0.77210	1.29
Eutrophication potential (EP)	kg PO_4^{3-}	0.00027	23.01	0.00042	11.02
Fresh water aquatic ecotoxicity (FWAE)	kg 1,4-DB [z]	0.01934	4.35	0.03490	3.48
Human toxicity (HT)	kg 1,4-DB	0.04410	15.60	0.10200	1.03
Terrestrial ecotoxicity (TE)	kg 1,4-DB	0.00066	4.63	0.00144	0.79

[z] 1,4-dichlorobenzene (DB).

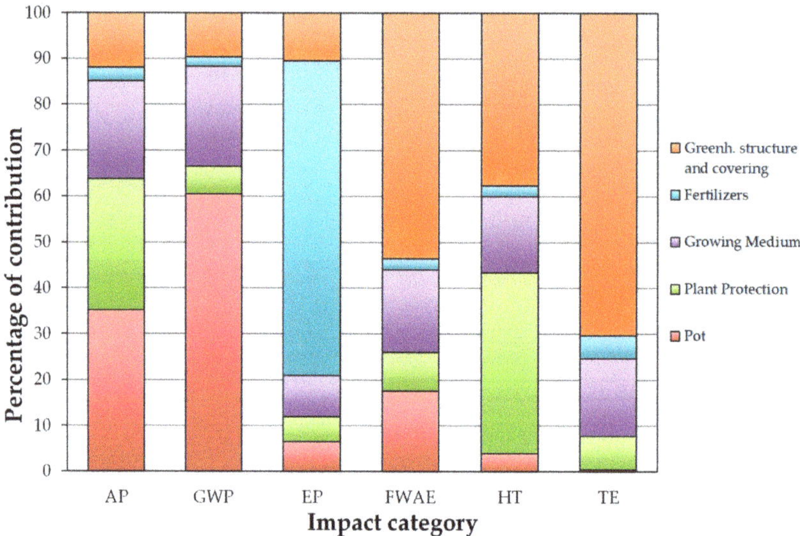

Figure 1. Relative contribution of different inputs for cyclamen potted plant production. The impact categories assessed are: Acidification potential (AP), global warming potential (time horizon of 100 years) (GWP), eutrophication potential (EP), fresh water aquatic ecotoxicity (FWAE), human toxicity (HT) and terrestrial ecotoxicity (TE).

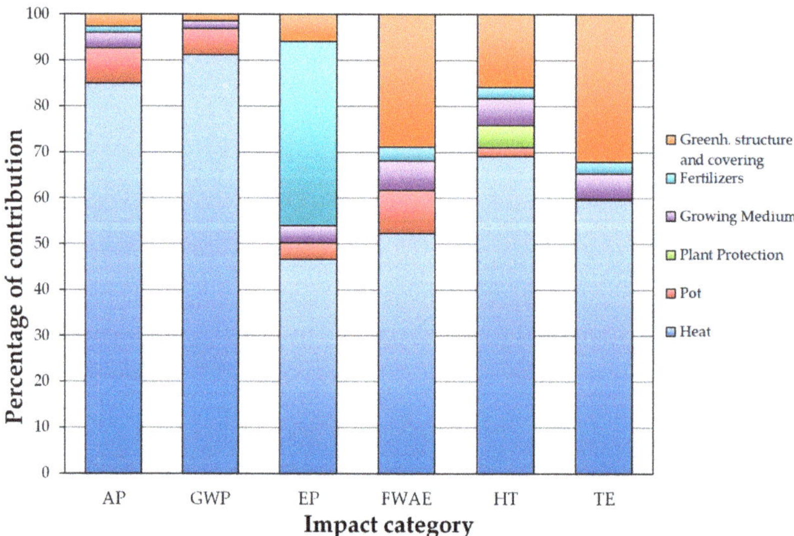

Figure 2. Relative contribution of different inputs for zonal geranium potted plant production. The impact categories assessed are: AP, GWP, EP, FWAE, HT and TE.

Relative contributions in the impact categories are depicted graphically in Figures 1 and 2. The reported percentages refer to average sample values. The contribution of some materials or structures showed little variation, given the relative uniformity of supply chain and input choices among the growers. Other inputs with less standardization showed significant differences in their contribution to impact categories, which will be discussed in the following paragraphs.

3.1. Cyclamen

Plastic containers were the major contributor (60.5%) for the GWP category, but also accounted for a significant share of impacts in AP (35.2%) and FWAE (17.6%) (Figure 1). All burdens were associated with material production, since no emissions were considered for use and end-of-life phases. Growing media components had an important share of impacts in the AP (21.3%), GWP (21.6%), FWAE (18%), HT (16.5%) and TE (17%) categories. Expanded perlite production and disposal was an important source of emissions for HT, TE and FWAE; emissions related to peat roadway transport from Baltic countries contributed mainly to GWP and AP categories (Figure 1). Greenhouse structure shared major burdens in FWAE (53.5%), HT (37.6%) and TE (70.4%) categories, mostly linked to production and disposal of steel frame and electricity consumption. Emissions related to production and use of plant protection products mainly influenced HT (39.4%) and AP (28.6%) categories; depending on chemical products type and frequency of treatments their contribution varied between 35.6% and 22.7% for AP, and between 43.8% and 32.5% for HT (Figure 1). Emissions related to fertilizer and water use contributed mainly (68.5%) to EP category results. The release of nitrate and phosphate in ground and surface water was directly linked to fertigation method and discharge mode and rate of nutrient solutions; the overall contribution of this phase varied between 44.6% for closed systems with no overhead application to 72% for open systems with frequent overhead applications (Figure 1). Fertigation management of cyclamen plants with the latter method was prevalent among the interviewed growers. Most studies on the environmental impact of potted plants have focused mainly on climate change (GWP) [38,39], while few studies conducted complete LCIA including other impact categories [34,37]. In accordance with our results, when referring to unheated crops with no artificial lighting, factors influencing GWP are mainly linked to manufacturing of plastic materials (containers and greenhouse cover) and growing media components (peat and expanded perlite). Fertilizer contribution to the EP category on the overall production process of cyclamen potted plants was also highlighted by Russo and De Lucia Zeller [37]. Their finding is in line with our results, suggesting that management practices aimed at reducing fertilizer use and leaching have the best chances for impact reduction in this category. The significant contribution of greenhouse structures to TE and FWAE categories is in line with similar studies on ornamental productions [34].

3.2. Zonal Geranium

Emissions deriving from production and use of diesel fuel burned to heat the greenhouse contributed a major share of impacts in all considered categories, accounting for over 91.3% of overall emissions in GWP and 84.7% in AP (Figure 2). Production and disposal of greenhouse frames contributed significantly to FWAE (28.7%), HT (17.9%) and TE (32.1%) categories (Figure 2). Fertilizer and water use contributed 40% of the impacts in the EP category. Since zonal geranium is often fertigated with ebb and flow systems, which allow for a reduction of direct emissions of both water and fertilizers, the contribution of this step was less variable than in cyclamen and ranged from 36.4% to 43.9% (Figure 2). Plastic pot contribution averaged 9.7% for FWAE, 7.7% for AP and 5.65% for GWP categories. The share of environmental burden from application of plant protection products and fertigation was not relevant for the selected impact categories, except for HT (4.8%) (Figure 2). These results are in line with other studies on protected crops that require energy inputs to actively control the greenhouse environment (light, temperature) or for preservation purposes [40]; the overall impact dramatically increases [34] and is almost entirely attributable to energy demand, as in the case of zonal geranium.

3.3. Effect of Alternative Practices on Cyclamen and Zonal Geranium Impact Assessment Results

Sensitivity analysis is a tool for studying the variability of LCIA results to input parameters and data. In the following sections we use it to assess the effects of different scenarios (management choices) on the environmental profile of our functional units. Since the chosen unit was a single potted plant, absolute impact values and variations observed in the analysis were extremely small. For this reason, the relevant differences are expressed in percentage on the impact potential.

Table 3 shows the results for the chosen categories of average production practices and for the alternative scenarios for cyclamen plants, highlighting the achievable impact. The reduction in chemical inputs attained through the application of integrated pest and pathogen management programs for cyclamen plants resulted in an overall reduction of potential impacts, which is relevant in particular for HT (−25%) and AP (−16.3%) categories (IPM in Table 3 vs. actual scenario in Table 2). For HT, this result was due primarily to reduction of soil emissions and manufacturing of active ingredients with fungicide activity, achieved through application of biological control agents and careful fertigation management.

Table 3. Sensitivity analysis for one cyclamen plant subjected to alternative practices. In relation to garden waste compost addition to the growing medium in option 1, NPK content of compost does not replace fertilizers and in option 2 NPK content replaces part of the mineral fertilizers applied through fertigation. IPM = integrated pest management.

Impact Category	Reference Unit	Compost		
	Equivalents	Option 1	Option 2	IPM
Acidification potential (AP)	kg SO_2	0.00035	0.00034	0.00030
Global warming potential (GWP)	kg CO_2	0.06980	0.06910	0.07260
Eutrophication potential (EP)	kg PO_4^{3-}	0.00027	0.00022	0.00027
Fresh water aquatic ecotoxicity (FWAE)	kg 1,4-DB [z]	0.01810	0.01780	0.01850
Human toxicity (HT)	kg 1,4-DB	0.04180	0.04100	0.03290
Terrestrial ecotoxicity (TE)	kg 1,4-DB	0.00063	0.00063	0.00064

[z] 1,4-dichlorobenzene (DB).

Use of compost as growing media component without changes in fertilizer application rate (option 1) showed relatively small further reduction potential compared to the option in which fertilization was also considered (option 2), linked mostly to reduced peat extraction and transport. Another study in which the environmental aspects of compost substitution was assessed [46] reported lower impact values for different categories, including climate change (another expression used for GWP), acidification potential, eutrophication potential and photochemical ozone formation. In this study, leaching tests for soil application suggested a potential higher impact of composts when considering potential impacts on human toxicity via water and soil, because of high release rates of heavy metals. These considerations partly support our results, since application of compost, that substitutes a 20% volume of peat in the growth medium, results in a slight reduction of several indicators, including GWP, that was reduced by only 7.4% and 6.4% in options 2 and 1, respectively. However, the reduction achieved by this practice had a limited relevance on the overall impact of the functional units. This can be explained by the small amount of peat replaced, the relative importance of growing media components in the assessed categories, and finally because of the impacts related to the compost production process. When considering also nutrient release from the compost amendment and subsequent reduction of fertigation needs, a significant reduction for the EP category (−19.6%) was observed, which can be explained both by reduction of fertilizer production and decreased leaching. We highlight that the minimum value of EP observed for cyclamen was very similar to that obtained for this scenario. This result is justified by data on cultivation with closed-loop fertigation systems with nutrient solution recirculation. To maximize impact reduction from nutrient production and leaching to surface and groundwater, a combination of fertigation management and use of nutrient-rich amendments in the growth medium could be a useful indication for best management practices.

Table 4 shows the impact of average production practices and the alternative scenarios for zonal geranium plants. We highlight how the potential for impact reduction was strongly limited by the major burdens linked to heating in all impact categories. Application of IPM programs achieved a moderate reduction of results for HT (−2.4%) category. Use of compost, not considering nutrient supply, achieved a reduction exceeding 1% of impact results for only FWAE (1.08%), TE (1.11%) and HT (1.63%) categories.

Table 4. Sensitivity analysis for one zonal geranium plant subjected to alternative practices. In relation to garden waste compost addition to the growing medium in option 1, NPK content of compost does not replace fertilizers and in option 2 NPK content replaces part of the mineral fertilizers applied through fertigation.

Impact Category	Reference Unit	Compost		
	Equivalents	Option 1	Option 2	IPM
Acidification potential (AP)	kg SO_2	0.00035	0.00034	0.00030
Global warming potential (GWP)	kg CO_2	0.77110	0.77090	0.77190
Eutrophication potential (EP)	kg PO_4^{3-}	0.00042	0.00036	0.00042
Fresh water aquatic ecotoxicity (FWAE)	kg 1,4-DB [z]	0.03450	0.03410	0.03510
Human toxicity (HT)	kg 1,4-DB	0.09990	0.09930	0.09910
Terrestrial ecotoxicity (TE)	kg 1,4-DB	0.00142	0.00139	0.00142

[z] 1,4-dichlorobenzene (DB).

When considering mineral fertilizing offsets, the differences increased, in particular for EP that shows a 14% reduction in the final result. This value was lower than the observed minimum, highlighting the higher uniformity and technological level adopted for zonal geranium fertigation. In a trial on geranium bedding plants [20], compost from selected materials had a supporting effect on growth of geranium plants, providing an increased nutrient budget in the growing media and an increased uptake and nutrient content in plant tissues. The use of peat-free substrate increases production risk and requires expertise, and often alternative substrates cannot be adopted [36]; however, the addition of compost to growing media for geranium growth may be increased to 40%, providing a large part of its nutrient requirements, as evidenced by Perner et al. [44] in growth trials conducted with potted geranium. The adoption of this practice therefore shows a potential for impact reduction in the EP category, if mineral fertilizer inputs are accordingly reduced. The alternative practice we investigated falls among the priorities in pollution prevention listed as Best Agricultural Practices for protected crops in Mediterranean Climates [47], yet their improvement potential differs greatly depending on the set of impact categories and technological level, material and energy requirements of the investigated production system. For low energy-input crops such as cyclamen, the decrease in fertilizer and pesticide use can result in a significant impact reduction for most of the selected categories. The potential benefit resulting from combined application was 32% for HT, 20% for AP and EP, 12.5% for FWAE and 10% for GWP.

For zonal geranium, we highlight how reduction of energy input is the first priority for soilless heated crops, since best practices for other highly impacting materials (plastic containers and cover) have already been adopted. The reduced amount of fertilizer and plant protection product translates to a relatively irrelevant contribution, except for the EP category.

4. Conclusions

In this study we investigated the environmental impact aspects of the cultivation of cyclamen and zonal geranium starting from data coming from different greenhouse farms located in the Treviso province of Italy. Given the fragmented structure of the production chain for floriculture products in this region, the definition of common practices and their characterization should be linked to a variability measure in order to include the complexity and plurality of structures and management choices in the final results. In the case of cyclamen production, technological level and management choices can greatly affect the values obtained for different environmental indicators, in particular with regard to fertigation management and use of plant protection products. The results of the analysis also highlighted how the efficiency of reduction measures should always be checked with a life cycle study on the production or process to address (e.g., potted ornamental plants).

While "sustainable" choices such as composting and reuse of waste biomass and reduction of chemical treatments have a significant benefit when applied to crops grown in a passive greenhouse, energy saving and changes in fuel type should be the main concern when aiming to reduce the impacts for crops requiring active control of the growth environment, as in the case of zonal geranium.

Supplementary Materials: The following are available online at https://www.mdpi.com/2311-7524/7/1/8/s1, questionnaire.

Author Contributions: Conceptualization, J.E.B., P.S. and G.Z.; methodology, J.E.B. and L.C.; formal analysis, J.E.B., L.C. and G.Z.; investigation, J.E.B., C.N. and L.C.; resources, C.N. and G.Z.; data curation, J.E.B., C.N. and G.Z.; writing—original draft, J.E.B.; writing—review and editing, P.S., C.N. and G.Z.; visualization, J.E.B., L.C. and P.S.; supervision, G.Z.; funding acquisition, P.S. All authors have read and agreed to the published version of the manuscript.

Funding: This research was partially funded by Measure 124 of the Rural Development Program 2007–2013 of the Veneto Region (Italy), Project "REFF".

Institutional Review Board Statement: Not applicable.

Informed Consent Statement: Not applicable.

Data Availability Statement: Data available on request due to privacy restrictions.

Acknowledgments: The authors are grateful to the Florveneto association and to the 20 farmers who participated in this study.

Conflicts of Interest: The authors declare no conflict of interest.

References

1. AIPH; Union Fleurs. *International Statistics Flowers and Plants 2018*; International Association of Horticultural Producers Horticulture House: Chilton, UK, 2018; Volume 66, p. 204.
2. Veneto Agricoltura 2016. Andamento Congiunturale del Comparto Florovivaistico. Bollettino n. 32. Available online: https://www.venetoagricoltura.org/wp-content/uploads/2019/03/Bollettino-FV-n_32.pdf (accessed on 18 December 2020).
3. Evers, B.J.; Amoding, F.; Krishnan, A. Social and Economic Upgrading in Floriculture Global Value Chains: Flowers and Cuttings GVCs in Uganda. *SSRN J.* **2014**, *42*. Available online: https://papers.ssrn.com/sol3/papers.cfm?abstract_id=2456600 (accessed on 20 October 2020). [CrossRef]
4. Marble, S.C.; Prior, S.A.; Runion, G.B.; Torbert, H.A.; Gilliam, C.H.; Fain, G.B. The Importance of Determining Carbon Sequestration and Greenhouse Gas Mitigation Potential in Ornamental Horticulture. *Horts* **2011**, *46*, 240–244. [CrossRef]
5. UNEP (United Nations Environmental Program). *International Declaration on Cleaner Production*; Division of Technology, Industry and Economics: Paris, France, 1998.
6. de Moraes, G.J.; Tamai, M.A. Biological Control Control of Tetranychu Spp. on Ornamental Plants. *Acta Hortic.* **1999**, *482*, 247–252. [CrossRef]
7. Rose, S.; Yip, R.; Punja, Z.K. Biological Control of Fusarium and Pythium Root Rots on Greenhouse Cucumbers Grown in Rockwool. *Acta Hortic.* **2004**, *635*, 73–78. [CrossRef]
8. Minuto, A.; Grasso, V.; Gullino, M.L.; Garibaldi, A. Chemical, Non-Chemical and Biological Control of Phytophthora Cryptogea on Soilless-Grown Gerbera. *Acta Hortic.* **2005**, *698*, 153–158. [CrossRef]
9. Ślusarski, C. Evaluation of Chemical and Biological Control Methods for Their Potential to Reduce Bacterial Canker of Tomato in a Greenhouse Stonewool Cultivation System. *Acta Hortic.* **2005**, *698*, 299–304. [CrossRef]
10. Penningsfeld, F. Use of Slow-Release Fetilizers in Peat Substrates. *Acta Hortic.* **1975**, *50*, 125–130. [CrossRef]
11. Kobel, F. The Use Slow Release Fertilizers in Pot Plant Production. *Acta Hortic.* **1975**, *50*, 131–134. [CrossRef]
12. Markus, D.K.; Flannery, R.L. Macronutrient Status in Potting Mix Substrates and in Tissue of Azaleas from the Effects of Slow-Release Fertilizer. *Acta Hortic.* **1983**, *133*, 179–190. [CrossRef]
13. Nicese, F.P.; Ferrini, F. Verifica della possibilità di uso di reflui industriali trattati per irrigazione di arbusti ornamentali in contenitore. *Italus Hortus* **2003**, *10* (Suppl. 4), 129–133.
14. Incrocci, L.; Pardossi, A.; Marzialetti, P. Innovazioni tecnologiche per l'irrigazione delle colture florovivaistiche. *Italus Hortus* **2001**, *11*, 43–51.
15. Darras, A.I. Implementation of Sustainable Practices to Ornamental Plant Cultivation Worldwide: A Critical Review. *Agronomy* **2020**, *10*, 1570. [CrossRef]
16. Castronuovo, D.; Picuno, P.; Manera, C.; Scopa, A.; Sofo, A.; Candido, V. Biodegradable Pots for Poinsettia Cultivation: Agronomic and Technical Traits. *Sci. Hortic.* **2015**, *197*, 150–156. [CrossRef]

17. Zanin, G.; Coletto, L.; Passoni, M.; Nicoletto, C.; Bonato, S.; Ponchia, G.; Sambo, P. Organic By-Product Substrate Components and Biodegradable Pots in the Production of *Pelargonium × hortorum* Bailey and *Euphorbia Pulcherrima* L. *Acta Hortic.* **2016**, *1112*, 371–378. [CrossRef]
18. Ponchia, G.; Passoni, M.; Bonato, S.; Nicoletto, C.; Sambo, P.; Zanin, G. Evaluation of Compost and Anaerobic Digestion Residues as a Component of Growing Media for Ornamental Shrub Production. *Acta Hortic.* **2017**, *1168*, 71–78. [CrossRef]
19. Ribeiro, H. Fertilisation of Potted Geranium with a Municipal Solid Waste Compost. *Bioresour. Technol.* **2000**, *73*, 247–249. [CrossRef]
20. Massa, D.; Malorgio, F.; Lazzereschi, S.; Carmassi, G.; Prisa, D.; Burchi, G. Evaluation of Two Green Composts for Peat Substitution in Geranium (*Pelargonium Zonale* L.) Cultivation: Effect on Plant Growth, Quality, Nutrition, and Photosynthesis. *Sci. Hortic.* **2018**, *228*, 213–221. [CrossRef]
21. Bassan, A.; Bona, S.; Nicoletto, C.; Sambo, P.; Zanin, G. Rice Hulls and Anaerobic Digestion Residues as Substrate Components for Potted Production of Geranium and Rose. *Agronomy* **2020**, *10*, 950. [CrossRef]
22. Gong, X.; Li, S.; Sun, X.; Wang, L.; Cai, L.; Zhang, J.; Wei, L. Green Waste Compost and Vermicompost as Peat Substitutes in Growing Media for Geranium (*Pelargonium Zonale* L.) and Calendula (*Calendula Officinalis* L.). *Sci. Hortic.* **2018**, *236*, 186–191. [CrossRef]
23. Massa, D.; Prisa, D.; Lazzereschi, S.; Cacini, S.; Burchi, G. Heterogeneous Response of Two Bedding Plants to Peat Substitution by Two Green Composts. *Hort. Sci.* **2018**, *45*, 164–172. [CrossRef]
24. Bassan, A.; Sambo, P.; Zanin, G.; Evans, M.R. Use of Fresh Rice Hulls and Anaerobic Digestion Resedues as Substrates Alternative to Peat. *Acta Hortic.* **2012**, *927*, 1003–1010. [CrossRef]
25. Bot, G.P.A. The Solar Greenhouse; Thechnology for Low Energy Cnsumption. *Acta Hortic.* **2004**, *633*, 29–33. [CrossRef]
26. Bakker, J.C.; Adams, S.R.; Boulard, T.; Montero, J.I. Innovative Technologies for an Efficient Use of Energy. *Acta Hortic.* **2008**, *801*, 49–62. [CrossRef]
27. Dieleman, J.A.; de Visser, P.H.B.; Vermeulen, P.C.M. Reducing the Carbon Footprint of Greenhouse Grown Crops: Re-Designing LED-Based Production Systems. *Acta Hortic.* **2016**, *1134*, 395–402. [CrossRef]
28. Baeza, E.J.; Pérez-Parra, J.; Montero, J.I. Effect of Ventilator Size on Natural Ventilation in Parral Greenhouse by Means of CFD Simulations. *Acta Hortic.* **2005**, *691*, 465–472. [CrossRef]
29. Molina-Aiz, F.D.; Valera, D.L.; Peña, A.A.; Gil, J.A. Optimisation of Almería-Type Greenhouse Ventilation Performance with Computational Fluid Dynamics. *Acta Hortic.* **2005**, *691*, 433–440. [CrossRef]
30. Valera, D.L.; Molina, F.D.; Alvarez, A.J.; López, J.A.; Terrés-Nicoli, J.M.; Madueño, A. Contribution to Characterisation of Insect-Proof Screens: Experimental Measurements in Wind Tunnel and CFD Simulation. *Acta Hortic.* **2005**, *691*, 441–448. [CrossRef]
31. Sase, S. Air Movement and Climate Uniformity in Ventilated Greenhouses. *Acta Hortic.* **2006**, *719*, 313–324. [CrossRef]
32. ISO 14040:2006. *ISO 14040: 2006 Environmental Management—Life Cycle Assessment—Principles and Framework*; BSI: London, UK, 2006.
33. Bonaguro, J.E.; Coletto, L.; Zanin, G. Environmental and Agronomic Performance of Fresh Rice Hulls Used as Growing Medium Component for *Cyclamen Persicum* L. Pot Plants. *J. Clean. Prod.* **2017**, *142*, 2125–2132. [CrossRef]
34. Bonaguro, J.E.; Coletto, L.; Samuele, B.; Zanin, G.; Sambo, P. Environmental Impact in Floriculture: LCA Approach at Farm Level. *Acta Hortic.* **2016**, *1112*, 419–424. [CrossRef]
35. Wandl, M.-T.; Haberl, H. Greenhouse Gas Emissions of Small Scale Ornamental Plant Production in Austria—A Case Study. *J. Clean. Prod.* **2017**, *141*, 1123–1133. [CrossRef]
36. Havardi-Burger, N.; Mempel, H.; Bitsch, V. Sustainability Challenges and Innovations in the Value Chain of Flowering Potted Plants for the German Market. *Sustainability* **2020**, *12*, 1905. [CrossRef]
37. Russo, G.; De Lucia Zeller, B. Environmental Evaluation by Means of LCA Regarding the Ornamental Nursery Production in Rose and Sowbread Greenhouse Cultivation. *Acta Hortic.* **2008**, *801*, 1597–1604. [CrossRef]
38. Koeser, A.K.; Lovell, S.T.; Petri, A.C.; Brumfield, R.G.; Stewart, J.R. Biocontainer Use in a *Petunia × hybrida* Greenhouse Production System: A Cradle-to-Gate Carbon Footprint Assessment of Secondary Impacts. *HortScience* **2014**, *49*, 265–271. [CrossRef]
39. Lazzerini, G.; Lucchetti, S.; Nicese, F.P. Green House Gases (GHG) Emissions from the Ornamental Plant Nursery Industry: A Life Cycle Assessment (LCA) Approach in a Nursery District in Central Italy. *J. Clean. Prod.* **2016**, *112*, 4022–4030. [CrossRef]
40. Abeliotis, K.; Barla, S.-A.; Detsis, V.; Malindretos, G. Life Cycle Assessment of Carnation Production in Greece. *J. Clean. Prod.* **2016**, *112*, 32–38. [CrossRef]
41. Heijungs, R.; Guinée, J.B.; Huppes, G.; Lamkreijer, R.M.; Udo de Haes, H.A.; Wegener Sleeswijk, A.; Ansems, A.M.M.; Eggels, P.G.; van Duin, R.; de Goede, H.P. *Environmental Life Cycle Assessment of Products. Guide (Part1) and Background (Part 2)*; CML Leiden University: Leiden, The Netherlands, 1992.
42. Prasuhn, V. *Erfassung der PO4-Austräge für die Ökobilanzierung—SALCA-Phosphor*; Agroscope FAL Reckenholz: Zürich, Switzerland, 2006.
43. European Parliament. Regulation (EC) No 1107/2009 Concerning the Placing of Plant Protection Products on the Market. *Off. J. Eur. Union* **2009**, *L309*, 1–50.
44. Perner, H.; Schwarz, D.; Bruns, C.; Mäder, P.; George, E. Effect of Arbuscular Mycorrhizal Colonization and Two Levels of Compost Supply on Nutrient Uptake and Flowering of Pelargonium Plants. *Mycorrhiza* **2007**, *17*, 469–474. [CrossRef]

45. Vecchietti, L.; De Lucia, B.; Russo, G.; Rea, E.; Leone, A. Environmental and Agronomic Evaluation of Containerized Substrates Developed from Sewage Sludge Compost for Ornamental Plant Production. *Acta Hortic.* **2013**, *1013*, 431–439. [CrossRef]
46. Boldrin, A.; Hartling, K.R.; Laugen, M.; Christensen, T.H. Environmental Inventory Modelling of the Use of Compost and Peat in Growth Media Preparation. *Resour. Conserv. Recycl.* **2010**, *54*, 1250–1260. [CrossRef]
47. Baudion, W.; FAO. *Good Agricultural Practices for Greenhouse Vegetable Crops: Principles for Mediterranean Climate Areas*; FAO Plant Production and Protection Paper; Food and Agricultural Organization of the United Nations (FAO): Rome, Italy, 2013; ISBN 978-92-5-107649-1.

MDPI

Review

Overview of the Dynamic Role of Specialty Cut Flowers in the International Cut Flower Market

Anastasios Darras

Laboratory of Floriculture and Landscape Architecture, Department of Agriculture, University of Peloponnese, 24100 Kalamata, Greece; a.darras@uop.gr; Tel.: +30-27210-45199

Abstract: The global cut flower industry has faced serious challenges over the years, but still remains an important sector of agriculture. Floriculture businesses seek new, innovative trends and niches to help increase product sales. Specialty cut flower (SCF) production has increased in the past 20 years in the US, Australia, Africa, and Europe. SCF production and sales could increase further if these new products were supported by dynamic marketing campaigns that focus on their strengths compared to the traditional cut flowers (TCF) such as roses, carnations, gerberas, and chrysanthemums. The major strength of SCF is the eco-friendly profile, which is associated to low CO_2 footprints and environmental outputs. This contrasts TCF cultivation, which is associated to high energy inputs, especially at the traditional production centres (e.g., The Netherlands). It is suggested that environmental legislations, production costs, and customer demand for eco-friendly products will positively affect future SCF cultivation and sale.

Keywords: roses; gerberas; chrysanthemums; sustainability; floriculture; environmental impact; CO_2 footprint

Citation: Darras, A. Overview of the Dynamic Role of Specialty Cut Flowers in the International Cut Flower Market. *Horticulturae* **2021**, *7*, 51. https://doi.org/10.3390/horticulturae7030051

Academic Editor: Piotr Salachna

Received: 22 February 2021
Accepted: 12 March 2021
Published: 14 March 2021

1. Introduction

Global cut flower production and consumption has overcome serious challenges in the past 20 years, especially those related to global economic recessions. The EU holds the first place in cut flower and ornamental potted plants sales with 31.0% of the global value, with China and the USA in second and third place, holding 18.6% and 12.5%, respectively [1]. Within the EU, in 2016, the Netherlands had the most sales of cut flowers and ornamental plants, with France and Italy in second and third places, respectively [1]. Cut flower and ornamental plant sales in the EU increased by 7% (approx. 1.4 billion euro) from 2006 to 2016, indicating a slow, but steady increase, despite the elaborate global economic status [1]. On the contrary, the number of cut flower producers in the USA declined significantly from 2007 to 2015 [2], with many of them forced out from the floriculture industry during the 2008–2009 recessionary shakeout period [3]. Cut flower production in the USA showed a modest increase from 2015 to 2018 [4,5]. The reductions recorded by the USDA between 2007 and 2015 reached 30%, indicating that cut flower production has shifted to new worldwide players such as China, Colombia, and Ecuador [6]. In 2017, China came first in sales of cut flowers and first in cut flower exports to the EU.

The aim of this review was to analyse the dynamic role of specialty cut flowers (SCF) in a market overwhelmed by traditional cut flowers (TCF) holding the majority of sales. A critical analysis provides evidence that SCFs might serve as the environmentally friendly alternatives to TCFs and claim future higher production volumes and market shares globally.

2. Exploitation of the Endemic Flora by the Floriculture Sector

New specialty ornamental crops and cut flowers are often introduced from the endemic flora. Native species leaving their natural environment may become global market trends. More than 6000 species native to Asia, Europe, North America, South America,

Australia, Africa, and New Zealand were found in gardens in the USA [7]. The migration of ornamental plants has a deep historic background (18th century) and is still active to the present day.

Many South African native plants such as agapanthus (*Agapanthus africanus*), gerbera (*Gerbera jamesonii*), gladiolus (*Gladiolus* hybrids), gloriosa (*Gloriosa rothschildiana*), freesia (*Freesia hybrida*), leucadendron (*Leucadendron* sp.), leucospermum (*Leucospermum* sp.), ornithogalum (*Ornithogalum arabicum*), and protea (*Protea* sp.), were among the best-selling cut flower species during the previous decades [8]. Most of them are now considered as TCFs of great commercial success and are cultivated globally all-year round. They have been crossed or hybridised to produce numerous cultivars which gained commercial success [8]. The modern-day interest for South African native species lies on the breeding programs and/or on adaptations to various growing conditions. South Africa is regarded as a "hotspot" of diversity and an important source of new ornamental plants with the potential for commercial exploitation [8].

Cunningham et al. [9] presented many vital steps as part of a strategy to address and overcome serious challenges in the production of endemic cut flowers in Australia. Issues addressed were: (a) The implementation of existing national and international protocols for genetic resources protection, (b) the development of cultural branding and certification as marketing tools, (c) building business expertise and producer associations, and (d) increasing the reliability of supply chain of species such as *Geraldton waxflower*, *Anigozanthos* sp. *Boronia heterophylla*, *Leptospermum* sp., and *Grevillea* sp. [9]. In order to increase the economic and social impact of the native Australian SCF production and sale, local growers must improve their business and marketing skills, induce collaboration, exploit communication technologies, and increase product reliability and branding, while ensuring governmental support on protection of the initial genetic material. Similar challenges should be addressed by the South African floriculture markets [10]. The political isolation, the lack of complete distribution channels, the reduced quality compared to high grade European products, and the lack of export orientation were considered as the main barriers to valorisation of the South African native SCF. Although there is a dynamic increase in SCF production in the African continent, the labour organisations need to set labour standards for workers in production facilities [11].

3. Cultivation of TCF and SCF in a Globalised Market

The definition of SCF cannot be clear and precise, although scientists have tried to introduce a terminology that includes the following:

(a) SCF are annual or perennial species. Ornamental brunches from shrubs and trees are also included in the SCF group;
(b) they are categorized as species. Only a few of them have been hybridised to produce new varieties;
(c) they are seasonally produced in small quantities;
(d) they are mainly sold in local flower markets;
(e) they can be stored only for short periods of time;
(f) the majority of species is produced outdoors.

All-year round, intensive greenhouse cultivation of cut flowering stems may separate SCF from TCF. Species such as roses, gerberas, carnations, chrysanthemums, freesias, gypsophylla, eustoma, alstroemeria, phalaenopsis, anthurium etc. all fall into the TCF group (Table 1). The main differences between TCF vs. SCF could be related to:

(a) All year round vs. seasonally grown;
(b) numerous hybrids vs. original species;
(c) large stem numbers vs. small quantities;
(d) global sales vs. local markets;
(e) longer storage periods vs. short-period or no storage;
(f) greenhouse grown vs. outdoor cultivation.

Table 1. List of selected annual, perennial, and bulb specialty cut flowers (SCF) grown mainly outdoors at certain times of the year and traditional cut flowers (TCF) cultivated all-year round mainly inside greenhouses at different parts of the world [8,12–14].

SCF				TCF
Acacia L. species	*Campanula* L. species	*Forsythia x intermedia* Zab.	*Paeonia lactiflora* hybrids	*Alstroemeria* L. hybrids
Achillea L. species	*Capsicum annuum* L.	*Gaillardia x grandiflora* Van Houtte.	*Physalis alkekengi* L.	*Anthurium andraeanum* Andre
Agapanthus praecox Willd.	*Caryopteris x clandonensis* A. Simmonds	*Gomphrena globosa* L.	*Physostegia virginiana* (L.) Benth.	*Antirrhinum majus* L.
Ageratum houstonianum Mill.	*Centaurea cyanus* L.	*Helianthus annuus* L.	*Polyanthes tuberosa* L.	*Aster novi-belgii* L.
Allium L. species	*Chamelaucium uncinatum* Schauer	*Helichrysum bracteatum* Vent.	*Protea* R. Br.	*Celosia argentea* L.
Amaranthus caudatus L.	*Chimonanthus praecox* (L.) Link.	*Heliconia* angusta Vell., *H. aurantica* Ghiesbr.	*Ranunculus asiaticus* L.	*Dahlia* Kav. hybrids
Amaryllis belladonna L.	*Clematis lanuginosa* Lindl.	*Helleborus niger* L.	*Rudbeckia hirta* L.	*Dedranthema x grandiflorum* Kitam.
Ammi majus L.	*Cirsium japonicum* DC.	*Hippeastrum* Herb. hybrids	*Salix* sp.	*Delphinium x cultorum* Voss.
Anemone coronaria L.	*Consolida ambigua* L. P.W. Ball & Hey W.	*Hyacinthus orientalis* L.	*Salvia splendens* Sell & Roem. & Schult., *S. viridis* L.	*Dianthus caryophyllus* L.
Anigozanthos Labill. hybrids	*Convalaria majalis* L.	*Hydrangea macrophyla* Thunb.	*Scabiosa atropurpurea* L.	*Eustoma grandiflorum* Shinn
Aquilegia L. hybrids	*Cornus alba* L.	*Iris hollandica* Hort.	*Schinus molle* L.	*Freesia x hybrida* Bailey
Argeranthemum frutescens L. Schultz-Bip.	*Cosmos bipinnatus* Cav.	*Kniphofia uvaria* (L.) Oken.	*Scilla sibirica* L.	*Gerbera jamesonii* Bol.
Artemisia abrotanum L.	*Cotinus coggygria* Scop.	*Lathyrus odoratus* L.	*Skimmia* sp. L.	*Gladiolus* L. hybrids
Asclepias tuberosa L.	*Craspedia globosa* Benth.	*Lavatera trimestris* L.	*Strelitzia reginae* Banks.	*Gypsophila paniculata* L., *G. elegans* Bieb.
Astilbe x arendsii Arends.	*Crocosmia x crocosmiflora* Burb. & Dean	*Leucandedron* R. Br	*Syringa vulgaris* L.	*Lilium* L. hybrids (asiatic and oriental)
Astrantia major L.	*Cytisus canariensis* L.	*Leucospermum* R. Br.	*Telopea speciosissima* R. Br.	Orchidaceae (*Cattleya* Lindl., *Cymbidium* Sw., *Dendrobium* Sw.a dn *Phalaenopsis* Pfitz.)
Atriplex lumex	*Cynara scolymus* L.	*Liatris spicata* L. Wild.	*Trachelium caeruleum* Graham.	*Rosa* L. hybrids
Banksia sp. L.	*Digitalis purpurea* L.	*Limonium sinuatum* L. Mill.	*Tulipa gesneriana* L.	*Solidago* L. species
Boronia heterophylla L.	*Echinacea angustifolia* L. Moench.	*Lobelia* L. species	*Verbena bonariensis* L.	
Bouvardia sp.	*Echinops banaticus* Rochel.	*Matthiola incana* L. R. Br.	*Veronica longifolia* L.	
Brassica oleracea L.	*Eremurus* Bieb. species	*Nigela damascena* L.	*Veronicastrum virginicum* (L.) Farw.	
Buddleia davidii Franch., *B. globosa* L.	*Eryngium planum* L.	*Narcissus tazetta* L.	*Zantedeschia aethiopica* (L.) Spreng.	
Calendula officinalis L.	*Eucaris amazonica* Linden.	*Nerine bowdenii* (L.) Watson	*Zinnia elegans* Jacq.	
Callistephus chinensis L. Nees.	*Eucalyptus* L. species	*Ornithogalum arabicum* L.		

The above mentioned lists of SCF and TCF are dynamic and change over the years as a result of changing worldwide production and marketing strategies by companies in the floriculture sector. In this view, SCF cultivated all-year round due to their increased market demands may be considered as TCF. In a globalised market increased popularity may define the crop as SCF or TCF. Growers may cultivate SCF and TCF simultaneously to satisfy local and/or international market demands.

Cultivation of SCF may increase local growers' income [14–18]. Growers located in North America found great potential in growing "alternative forest crops" such as salix species (i.e., *Salix matsudana*, *S. caprea*, *S. purpurea*), cornus (i.e., *Cornus sericea*), forsythias (*Forsythia* sp.), and various other flowering bunches [17]. In the North America wholesale markets, more than 1 million bunches of curly and pussy willow were sold in 2001. An additional 152,000 forsythia stems and more than 140,000 *Cornus* sp. bunches were also sold [17]. Willow cut stem growers in North America had great experience on cultivation, but they were not aware or informed about fertilization, pest management, and postharvest handling [19]. More research is required on SCF production and postharvest handling to provide solutions to growers and sellers [4,19].

In Australia, SCF production involved a wide range of native species dominated by the waxflower (*Chamelaucium uncinatum*), the kangaroo paw (*Anigozanthos* spp.), and the thryptomene (*Thryptomene* spp.) [9]. At least 64 other countries produce endemic Australian cut flowers with Israel, USA (California), South Africa, Ecuador, and Colombia being among them. *C. uncinatum* is a fast growing, evergreen shrub cultivated outdoors that produces flowering stems during winter after going through the short-day autumn period [20]. Australian native acacia species such as *A. dealbata*, *A. retinodes*, and *A. baileyana* are grown commercially in Australia and in the Mediterranean (i.e., Italy, Israel) for their impressive inflorescences [9,21]. Many SCF such as *Antirrhinum majus*, *Echinacea purpurea*, *Helianthus annuus*, *Limonium sinuatum*, *Matthiola incana*, *Scabiosa atropurpurea*, *Zinnia elegans*, and many more have increased their share in the US market [14,15]. According to the study presented by Starman et al. [15], 16 out of the 19 field grown SCF crops were profitable for commercial cultivation. Grower's income from *Achillea filipedulina*, *Liatris spicata*, *Veronica spicata*, and *Centranthus ruber* production increased linearly with increasing price/bunch sales. Stem prices varied during the year, and generally peaked around major holidays [15]. Among those species, *Cosmos bippinatus* cut flowering stems gave the highest income to growers (e.g., $10.63–$13.62). *Zinnia elegans*, *Scabiosa atropurpurea*, and *Antirrhinum majus* were also profitable for growers with incomes ranging from $3.36 to $4.51 [15]. Byczynski [22] stated that SCF growers could profit $25,000 to $35,000 per year, per ha of cultivation. Locally cultivated *Helianthus annuus* "Firecracker" plants would potentially be sold to wholesalers, retailers, and consumers in prices similar to those of the imported *H. annuus* flowering stems [18].

Although there is great potential in SCF cultivation, TCF hold the major part of sales in the local and international markets. In FloraHolland, the largest market of floricultural products worldwide, roses and chrysanthemums are the first and second best-selling cut flowers, respectively. The tulips, the gerberas, and the liliums came 3rd, 4th, and 5th in the top-5 list of cut flowers sold in 2019 [23]. Tulips may be considered as the best-selling SCF in the world, produced mainly in the Netherlands and presented as their national species.

The SWOT analysis shown in Table 2 provides useful comparisons between SCF and TCF. Although, growers' and sellers' decisions on cultivation and trade are complex and are often related to several factors and idiosyncrasies such as social and environmental legislations, infrastructures, environmental conditions, labour, and transportation costs. In a constantly changing global market, the TCF cultivation is the assured solution for growers and sellers, which, however, shows weaknesses associated with environmental legislations, CO_2 footprints, pesticide residues, and increased energy demands (Table 2). On the contrary, production of SCF could serve as the new alternative choices for retailers and florists who always seek niche markets to sell their products [16]. This was the case for Oklahoma (US) ornamental-horticulture and cut flower retailers that indicated the positive outcome in retail by using a greater variety of species [16]. Local production for domestic markets could be another strength of SCF (Table 2). In the US, production of local SCF increased the past 20 years and challenged imports of TCF grown in South America or other locaters [4,24]. Consumers who bought locally grown products had the perception of benefiting the local economy [25].

Table 2. Strengths, weaknesses, opportunities, and threats (SWOT) of SCF and TCF cultivated worldwide.

Species	Strengths	Weaknesses	Opportunities	Threats
SCF	a. Low cost, mainly outdoor b. Environmental friendly cultivation with low inputs c. Sustainable and low outputs d. Local production e. Variety of aesthetic features f. Many SCF have low longevity	a. Lack of identification and recognition by the customers b. Low commercial value c. Postharvest characteristics (low vase life, short storage period etc)	a. Sustainable production, reduction of CO_2 footprint, minimum phytochemicals b. New trends for floriculture industry and for the retail c. Partial replacement of TCF	a. Quality and vase life b. Customer confusion by the large number of species
TCF	a. Recognizable worldwide b. Varieties with high postharvest longevity c. Most appropriate for postharvest handling and storage d. Associated with symbolic meanings and social events e. Many different varieties provide various aesthetic features f. New varieties are associated with new marketing trends g. New varieties can be cultivated in different environments	a. High production costs due to greenhouse construction and automations and energy inputs b. High environmental outputs and CO_2 footprints c. Pesticide residues	a. Production of new, less energy demanding varieties b. Use of renewable energy sources at cultivation	a. New environmental legislations may put restrictions to traditional cultivations

4. Sustainable Production of SCF vs. TCF

As the green industry continues to mature, differentiation is an increasingly important business strategy [26]. One way to accomplish this is by adopting environmentally friendly behaviours that will attract consumers with environmental awareness [27–29]. These potential consumers are more likely to purchase environmental friendly products with reduced CO_2 footprints [3,26]. There is a small, but considerable, percentage of people who were willing to pay more money for agricultural products associated with sustainable, eco-friendly cultivation procedures [25,30]. Mainstream consumers were willing to buy eco-friendly products, but only at a modest price difference. Special attention should be given to consumer education and other promotion-related programs based on partnership between universities and private bodies (i.e., Texas Superstar®) to increase sales of new cut ornamental products [31]. Growers were also willing to adopt eco-friendly practices in cultivation, although they were sceptical on the implementation within their current cultivation system [32]. Back in 2010, Dennis et al. [32] reported that none of the grower respondents in their survey were certified as sustainable.

The recognition of floricultural products as "sustainable" is complex and demanding. Sustainable production is achieved via the implementation of strict environmental and social protocols as defined by the national and international directives [11,33]. Restrictions on CO_2 footprints and global warming potential (GWP) may affect production of cut flowers in the future [33,34]. Over a public demand for cleaner agricultural products, the sustainable SCF cultivation may serve as the environmental friendly alternative option. This can be a major strength of SCF compared to TCF (Table 2). Wandl and Haberl [35] showed that summer and spring SCF grown outdoors had <0.1 kg $CO_{2\,eq}$, while rose cultivation produced up to 13-fold more kg CO_2. The main differences in CO_2 production between the SCF and TCF were associated with excessive heating and electricity use "off-season" (i.e., the cold days of the year). Life cycle assessments (LCA) showed that increased CO_2 outputs for the production of roses, chrysanthemums, and gerberas were profound in Central and Northern European countries such as the Netherlands, Germany, and Austria [35–38]. Significant differences in CO_2 footprints, acidification, global warming, human toxicity, marine ecotoxicity, terrestrial ecotoxicity, and phytochemical oxidation were reported for roses produced in Dutch greenhouses compared to those produced in Ecuadorian facilities [33,38]. While CO_2 for roses produced in Kenya and Ethiopia ranged between 0.4 and 3.7 kg $CO_{2\,eq}$, the Dutch roses released 16–29 kg $CO_{2\,eq}$ [37]. In Greece, the cultivation of carnations in non-heated greenhouses produced only 0.316 kg $CO_{2\,eq}$ [39], indicating that heating during winter is the single most important factor contributing to greenhouse gas (GHG) emissions. As a result, future environmental legislations will apply limitations to TCF cultivation at traditional production centres, or may help in shifting production of TCF to countries in the African continent or at South America [33,34,40–42].

Fertilization and agrochemical use both contribute to the environmental outputs during cultivation. Growers of TCF support the integrated nutrition management (INM) and integrated pest and disease management (IPDM) programs to reduce their environmental footprints [33]. However, pesticide residue levels in roses, gerberas, and chrysanthemums highly concerned authorities and consumers in the EU in the past decade. In a study conducted in Belgium, 107 active ingredients were detected in harvested rose, gerbera, and chrysanthemum bunches [43]. Among them, roses were the most contaminated flowers with 14 distinct substances detected per sample and a total concentration of 26 mg kg^{-1} for a single rose sample. Substances such as acephate, methiocarb, monocrotophos, methomyl, deltamethrin, etc., could generate direct toxic effects to the nervous system of florists and consumers [43]. No research was found on pesticide residues detected on SCF. Generally, SCF suffer fewer fungal and pest contaminations during production, and therefore require minimal amounts of phytochemicals. SCF crop rotation and seasonal production might be the key to less infections and herbivore attacks. In every case, the implementation of IPDM to ornamental crop production may eventually reduce pesticide residues and improve the profile of the floricultural products.

5. Postharvest Performance and Quality

Vase life (VL) of cut flowers is considered as the single most important factor affecting consumers' buying choices. The shorter the VL of flowers purchased, the lower the possibility for a repeated buying [44,45]. While TCF hybrids show exceptional VL, the SCF show shorter longevities [14]. Environment conditions and genotype were the most important preharvest factors that contributed to inflorescence VL [46]. For example, increased RH levels inside the greenhouse decreased stomatal conductance of rose plants during the day. It was shown that stomatal responsiveness could be improved by adjusting the humidity levels either during the day or at night. Mortensen and Gislerød [47] showed that severe drought stress during growth of roses at high RH increased their VL. Longevity of vars. "Akito" and "First Red" roses decreased in winter as a result of higher humidity levels inside the greenhouses [48].

Harvest time and stage of development significantly affected VL of TCF and SCF [49]. Harvest procedures vary among species. For example, species of the Asteraceae family (i.e., chrysanthemums, gerberas, zinnias, etc.) should be in full maturity, whereas inflorescences with multiple buds are harvested at 25–50% open flowers [12,14]. Harvesting the SCF inflorescences *Eremurus* "Line Dance" and "Tap Dance" at the stage of 0-florets open resulted in significantly longer VL compared to inflorescences harvested at 1–2 and at 3–5 floret rows open [50]. VL of cut *Capsicum* "Rio Light Orange" stems was affected by harvest stage with "partly mature fruit stage" being the best for harvest [51]. Both *Celosia argentea* and *Antirrhinum majus* inflorescences had maximum VL at early harvest stage of short head diameter and 0-florets open, respectively [52]. On the contrary, VL of *Leucocoryne coquimbensis* [53], *Viburnum tinus* [54], and *Spartium junceum* [55] inflorescences were not affected by the harvest stage, indicating that different responses are recorded among different genotypes. Stem length, leaf number, and harvest time significantly affected VL of various cut flower species [46]. Leaf number and stem length affected transpiration, water balance of cells, and water transport in xylem vessels, respectively. These were identified as the most crucial factors of flower wilting. Day-light and temperature during cultivation may also affect VL in response to stem hydraulic conductivity and carbohydrate storage [56,57].

Short-period storage and handling significantly affect VL of TCF and SCF [49]. Many SCF may respond well to low temperature storage [14]. SCF responded differently to postharvest handling (i.e., storage time and temperature, wet or dry storage). For example, wet or dry storage of *Eremurus* [50], *Matthiola incana* [58], *Capsicum* ornamental peppers [51], *Celosia argentea*, and *Antirrhinum majus* [52] was effective for up to 2 weeks. *Achillea filipendulina, Buddleia davidii, Cercis canadensis, Cosmos bipinnatus, Echinacea purpurea, Helianthus maximilianii, Penstemon digitalis*, and *Weigela* sp. were all positively affected by 1–2 week storage at temperatures ranging from 2.0 to 7.0 °C [59]. *S. junceum* inflorescences were successfully stored for 30 d at 3 °C, without any loss in VL [55]. Likewise, *Curcuma alismatifolia* was stored for 6 d at 7 °C without any loss in VL [60]. *Anigozanthos* spp. could be stored for up to 2 weeks at 4 °C without showing symptoms of wilt or petal discolouration [61].

The presence of ethylene inside storage facilities during handling or transport may severely reduce VL and quality of climacteric TCF and SCF. The SCF *Achillea filipendulina, Celosia argentea, Helianthus maximilianii, Penstemon digitalis*, and *Weigela* sp. were negatively affected by the presence of <1 μL L^{-1} ethylene inside the storage rooms [59]. Exposure of *Boronia heterophylla* flowers to 10 μL ethylene L^{-1} significantly reduced VL by 2.6 d compared to the un-exposed flowers [62]. *Curcuma alismatifolia* showed significant reductions in VL and bud opening after exposure to 0.5 - 2.0 μL L^{-1} ethylene [60]. *Spartium junceum* inflorescences suffered a detrimental flower and organ fall when exposed to 5 or 10 μL L^{-1} ethylene [55]. Treatments with 1-MCP, STS, Ag^{+}, and α-aminoisobutyric acid alleviated ethylene suffering and reduced aging symptoms. 1-MCP was found to be effective in various SCF including *Boronia heterophylla* [62], *Curcuma alismatifolia* [60], *Viburnum tinus* [54], *Celosia argentea, Antirrhinum majus* [52,63], *Eremurus* [50], and *Spartium junceum* [55]. 1-MCP increased the VL in ethylene-present or ethylene-free environments and delayed the aging process. Likewise, STS or Ag^{+} blocked ethylene-induced flower abscission and wilt and facilitated VL increases in most of the ethylene-sensitive SCF (i.e., *V. tinus, A. majus, B. heterophylla*).

6. Conclusions

Analysing the strengths and weaknesses of SCF and TCF, there are arguments over sustainability, marketing, and promotion of those cut flower groups. TCF are highly recognizable floricultural products, and biotechnology centres back-up their development with new varieties. On the contrary, SCF may be the eco-friendly alternative species to provide sustainable solutions to growers and consumers. SCF cultivation and sales have gained volumes over the last 15 years. SCF production can be sustainable with minimum energy and agrochemical inputs, although more research is required on VL extension and postharvest quality care.

Funding: This research received no external funding

Conflicts of Interest: The authors declare no conflict of interest.

References

1. Eurostat. Horticultural Products. Flowers and Ornamental Plants. Statistics 2006–2016. 2017. Available online: https://ec.europa.eu/info/sites/info/files/food-farming-fisheries/plants_and_plant_products/documents/flowers-ornamental-plants-statistics_en.pdf (accessed on 23 July 2020).
2. USDA 2016. Floriculture Crops 2015 Summary. USDA National Agricultural Statistics Service. Available online: https://downloads.usda.library.cornell.edu/usda-esmis/files/0p0966899/pz50gz655/8910jx14p/FlorCrop-04-26-2016.pdf (accessed on 23 July 2020).
3. Hall, C.R. Making cents of green industry economics. *HortTechnology* **2010**, *20*, 832–835. [CrossRef]
4. Loyola, C.E.; Dole, J.M.; Dunning, R. North American Specialty Cut Flower Production and Postharvest Survey. *HortTechnology* **2019**, *1*, 1–22. [CrossRef]
5. USDA 2019. Floriculture Crops 2018 Summary. USDA National Agricultural Statistics Service. Available online: https://www.nass.usda.gov/Publications/Todays_Reports/reports/floran19.pdf (accessed on 23 July 2020).
6. Eurostat. Horticultural Products. Flowers and Ornamental Plants. Statistics 2010–2019. 2018. Available online: https://ec.europa.eu/info/sites/info/files/food-farming-fisheries/plants_and_plant_products/documents/flowers-ornamental-plants-statistics_en.pdf (accessed on 23 July 2020).
7. Taylor, J. The migration of ornamental plants. *Chronica Hort.* **2010**, *50*, 21–24.
8. Reinten, E.Y.; Coetzee, J.H.; Van Wyk, B.E. The potential of South African indigenous plants for the international cut flower trade. *South. Afr. J. Bot.* **2011**, *77*, 934–946. [CrossRef]
9. Cunningham, A.B.; Garnett, S.T.; Gorman, J. Policy lessons from practice: Australian bush products for commercial markets. *GeoJournal* **2009**, *74*, 429. [CrossRef]
10. Van Rooyen, I.M.; Kirsten, J.F.; Van Rooyen, C.J.; Collins, R. The competitiveness of the South African and Australian flower industries: An application of three methodologies. In Proceedings of the 2001 45th AARES Conference, Adelaide, SA, Australia, 23–25 January 2001.
11. Riisgaard, L. Global value chains, labor organization and private social standards: Lessons from East African cut flower industries. *World Develop.* **2009**, *37*, 326–340. [CrossRef]
12. Dole, J.M.; Wilkins, H.F. *Floriculture: Principles and Species*; Pearson Prentice Hall: Upper Saddle River, NJ, USA, 2005.
13. Hunter, N. *The Art of Floral Design*; Delmar Thomson Learning: Florence, KY, USA, 2010.
14. Armitage, A.M.; Laushman, J.M. *Specialty Cut Flowers: The Production of Annuals, Perennials, Bulbs, and Woody Plants for Fresh and Dried Cut Flowers*; Timber Press: Portland, OR, USA, 2003.
15. Starman, T.W.; Cerny, T.A.; MacKenzie, A.J. Productivity and profitability of some field-grown specialty cut flowers. *HortScience* **1995**, *30*, 1217–1220. [CrossRef]
16. Dole, J.M.; Schnelle, M.A. A comparison of attitudes and practices among sectors of the Oklahoma floriculture industry. *HortTechnology* **1993**, *3*, 343–347. [CrossRef]
17. Josiah, S.J.; St-Pierre, R.; Brott, H.; Brandle, J. Productive conservation: Diversifying farm enterprises by producing specialty woody products in agroforestry systems. *J. Sustain. Agric.* **2008**, *23*, 93–108. [CrossRef]
18. Short, K.; Etheredge, C.L.; Waliczek, T.M. Studying the market potential for specialty cultivars of sunflower cut flowers. *HortTechnology* **2017**, *27*, 611–617. [CrossRef]
19. Saska, M.M.; Kuzovkina, Y.A.; Ricard, R.M. North American Willow Cut-stem Growers: A Survey of the Business Identities, Production Practices, and Prospective for the Crop. *HortTechnology* **2010**, *20*, 351–356. [CrossRef]
20. Shillo, R. Chamelaucium uncinatum. In *Handbook of Flowering*; Halevy, A., Ed.; CRC: Boca Raton, FL, USA, 1985; Volume 2, pp. 185–189.
21. Ratnayake, K.; Joyce, D. Native Australian Acacias: Unrealised ornamental potential. *Chron. Hortic.* **2010**, *50*, 19–22.
22. Byczynski, L. *The Flower Farmer: An Organic Grower's Guide to Raising and Selling Cut Flowers*, 2nd ed.; Chelsea Green Publishing: White River Junction, VT, USA, 2008.
23. Royal FloraHolland. Available online: www.floraholland.com (accessed on 15 September 2020).
24. Reid, M.S. Trends in flower marketing and postharvest handling in the United States. *Acta Hortic.* **2005**, *669*, 29–34. [CrossRef]
25. Khachatryan, H.; Rihn, A.; Campbell, B.; Behe, B.; Hall, C. How do consumer perceptions of "local" production benefits influence their visual attention to state marketing programs? *Agribusiness* **2018**, *34*, 390–406. [CrossRef]
26. Ingram, D.L.; Hall, C.R.; Knight, J. Understanding Carbon footprint in production and use of landscape plants. *HortTechnology* **2019**, *29*, 6–10. [CrossRef]
27. Oppenheim, P.P. Understanding the factors influencing consumer choice of cut flowers: A means-end approach. In Proceedings of the XIII International Symposium on Horticultural Economics, New Brunswick, NY, USA, 4–9 August 1996; Volume 429, pp. 415–422.
28. Behe, B.K.; Campbell, B.L.; Hall, C.R.; Khachatryan, H.; Dennis, J.H.; Yue, C. Consumer preferences for local and sustainable plant production characteristics. *HortScience* **2013**, *48*, 200–208. [CrossRef]
29. Campbell, B.; Khachatryan, H.; Behe, B.; Dennis, J.; Hall, C. Consumer perceptions of eco-friendly and sustainable terms. *Agric. Res. Econ. Rev.* **2015**, *44*, 21–34. [CrossRef]
30. Yue, C.; Hall, C. Traditional or specialty cut flowers? Estimating US consumers' choice of cut flowers at non-calendar occasions. *HortScience* **2010**, *45*, 382–386. [CrossRef]

31. Pemberton, B.; Arnold, M.; Davis, T.; Lineberger, D.; McKenney, C.; Rodriguez, D.; Stein, L.; Hall, C.; Palma, M.; De Los Santos, R. The Texas Superstar® Program: Success through Partnership. *HortTechnology* **2011**, *21*, 698–699. [CrossRef]
32. Dennis, J.H.; Lopez, R.G.; Behe, B.K.; Hall, C.R.; Yue, C.; Campbell, B.L. Sustainable production practices adopted by greenhouse and nursery plant growers. *HortScience* **2010**, *45*, 1232–1237. [CrossRef]
33. Darras, A.I. Implementation of sustainable practices to ornamental plant cultivation worldwide: A critical review. *Agronomy* **2020**, *10*, 1570. [CrossRef]
34. Kargbo, A.; Mao, J.; Wang, C.Y. The progress and issues in the Dutch, Chinese and Kenyan floriculture industries. *African J. Biotech.* **2010**, *9*, 7401–7408.
35. Wandl, M.T.; Haberl, H. Greenhouse gas emissions of small scale ornamental plant production in Austria—A case study. *J. Clean. Prod.* **2017**, *141*, 1123–1133. [CrossRef]
36. Williams, A. Comparative study of cut roses for the British market produced in Kenya and the Netherlands. In *Précis Report for World Flowers*; Cranfield University: Cranfield, UK, 2007; Volume 12, pp. 1–3.
37. Soode, E.; Lampert, P.; Weber-Blaschke, G.; Richter, K. Carbon footprints of the horticultural products strawberries, asparagus, roses and orchids in Germany. *J. Clean. Prod.* **2015**, *87*, 168–179. [CrossRef]
38. Franze, J.; Ciroth, A. A comparison of cut roses from Ecuador and the Netherlands. *Int. J. Life Cyc. Assess.* **2011**, *16*, 366–379. [CrossRef]
39. Abeliotis, K.; Barla, S.A.; Detsis, V.; Malindretos, G. Life cycle assessment of carnation production in Greece. *J. Clean. Prod.* **2016**, *112*, 32–38. [CrossRef]
40. Russo, G.; Buttol, P.; Tarantini, M. LCA (Life Cycle Assessment) of roses and cyclamens in greenhouse cultivation. *Acta Hortic.* **2008**, *801*, 359–366. [CrossRef]
41. Raynolds, L.T. Fair trade flowers: Global certification, environmental sustainability, and labor standards. *Rural Sociol.* **2012**, *77*, 493–519. [CrossRef]
42. Anonymous. *The EU Environmental Implementation Review 2019. Country Report, The Netherlands*; European Union: Luxembourg, 2019.
43. Toumi, K.; Vleminckx, C.; van Loco, J.; Schiffers, B. Pesticide residues on three cut flower species and potential exposure of florists in Belgium. *Int. J. Env. Res. Public Health* **2016**, *13*, 943. [CrossRef]
44. Rihn, A.L.; Yue, C.; Hall, C.; Behe, B.K. Consumer preferences for longevity information and guarantees on cut flower arrangements. *HortScience* **2014**, *49*, 769–778. [CrossRef]
45. Dennis, J.H.; Behe, B.K.; Fernandez, R.T.; Schutzki, R.; Page, T.J.; Spreng, R.A. Do plant guarantees matter? The role of satisfaction and regret when guarantees are present. *HortScience* **2004**, *40*, 142–145. [CrossRef]
46. Fanourakis, D.; Pieruschka, R.; Savvides, A.; Macnish, A.J.; Sarlikioti, V.; Woltering, E.J. Sources of vase life variation in cut roses: A review. *Postharvest Biol. Technol.* **2013**, *78*, 11–15. [CrossRef]
47. Mortensen, L.M.; Gislerød, H.R. Effect of air humidity variation on powdery mildew and keeping quality of cut roses. *Sci. Hort.* **2005**, *104*, 49–55. [CrossRef]
48. Pompodakis, N.E.; Terry, L.A.; Joyce, D.C.; Lydakis, D.E.; Papadimitriou, M.D. Effect of seasonal variation and storage temperature on leaf chlorophyll fluorescence and vase life of cut roses. *Postharvest Biol. Technol.* **2005**, *36*, 1–8. [CrossRef]
49. Joyce, D.C.; Faragher, J. Cut flowers. In *Crop Postharvest: Science and Technology. Pereshibles*; Rees, D., Farrell, G., Orchard, J., Eds.; Wiley-Blackwell: Ames, IA, USA, 2012; pp. 414–438.
50. Ahmad, I.; Dole, J.M.; Schiappacasse, F.; Saleem, M.; Manzano, E. Optimal postharvest handling protocols for cut 'Line Dance'and 'Tap Dance' Eremurus inflorescences. *Sci. Hort.* **2014**, *179*, 212–220. [CrossRef]
51. De FMFrança, C.; Dole, J.M.; Carlson, A.S.; Finger, F.L. Effect of postharvest handling procedures on cut Capsicum stems. *Sci. Hort.* **2017**, *220*, 310–316. [CrossRef]
52. Ahmad, I.; Dole, J.M. Optimal postharvest handling protocols for *Celosia argentea* var. *cristata* L. 'Fire Chief' and *Antirrhinum majus* L. 'Chantilly Yellow'. *Sci. Hort.* **2014**, *172*, 308–316.
53. Elgar, H.J.; Fulton, T.A.; Walton, E.F. Effect of harvest stage, storage and ethylene on the vase life of Leucocoryne. *Postharvest Biol. Technol.* **2003**, *27*, 213–217. [CrossRef]
54. Darras, A.I.; Akoumianaki-Ioannidou, A.; Pompodakis, N.E. Evaluation and improvement of post-harvest performance of cut *Viburnum tinus* inflorescence. *Sci. Hort.* **2010**, *124*, 376–380. [CrossRef]
55. Darras, A.I.; Kargakou, V. Postharvest physiology and handling of cut *Spartium junceum* inflorescences. *Sci. Hort.* **2019**, *252*, 130–137. [CrossRef]
56. Van Doorn, W.G.; Suiro, V. Relationship between cavitation and water uptake in rose stems. *Physiol. Plant.* **1996**, *96*, 305–311. [CrossRef]
57. Fanourakis, D.; Carvalho, S.M.; Almeida, D.P.; Heuvelink, E. Avoiding high relative air humidity during critical stages of leaf ontogeny is decisive for stomatal functioning. *Physiol. Plant.* **2011**, *142*, 274–286. [CrossRef] [PubMed]
58. Regan, E.M.; Dole, J.M. Postharvest handling procedures of *Matthiola incana* 'Vivas Blue'. *Postharvest Biol. Technol.* **2010**, *58*, 268–273. [CrossRef]
59. Redman, P.B.; Dole, J.M.; Maness, N.O.; Anderson, J.A. Postharvest handling of nine specialty cut flower species. *Sci. Hort.* **2002**, *92*, 293–303. [CrossRef]

60. Bunya-atichart, K.; Ketsa, S.; Van Doorn, W.G. Postharvest physiology of *Curcuma alismatifolia* flowers. *Postharvest Biol. Technol.* **2004**, *34*, 219–226. [CrossRef]
61. Joyce, D.C.; Shorter, A.J. Long term, low temperature storage injures kangaroo paw cut flowers. *Postharvest Biol. Technol.* **2000**, *20*, 203–206. [CrossRef]
62. Macnish, A.J.; Hofman, P.J.; Joyce, D.C.; Simons, D.H. Involvement of ethylene in postharvest senescence of *Boronia heterophylla* flowers. *Aust. J. Exp. Agric.* **1999**, *39*, 911–913. [CrossRef]
63. Çelikel, F.G.; Cevallos, J.C.; Reid, M.S. Temperature, ethylene and the postharvest performance of cut snapdragons (*Antirrhinum majus*). *Sci Hort.* **2010**, *125*, 429–433. [CrossRef]

MDPI
St. Alban-Anlage 66
4052 Basel
Switzerland
Tel. +41 61 683 77 34
Fax +41 61 302 89 18
www.mdpi.com

Horticulturae Editorial Office
E-mail: horticulturae@mdpi.com
www.mdpi.com/journal/horticulturae

www.ingramcontent.com/pod-product-compliance
Lightning Source LLC
LaVergne TN
LVHW070626170726
843515LV00004B/191
9783036546858